IMAGES
of America

NEW YORK MILLS

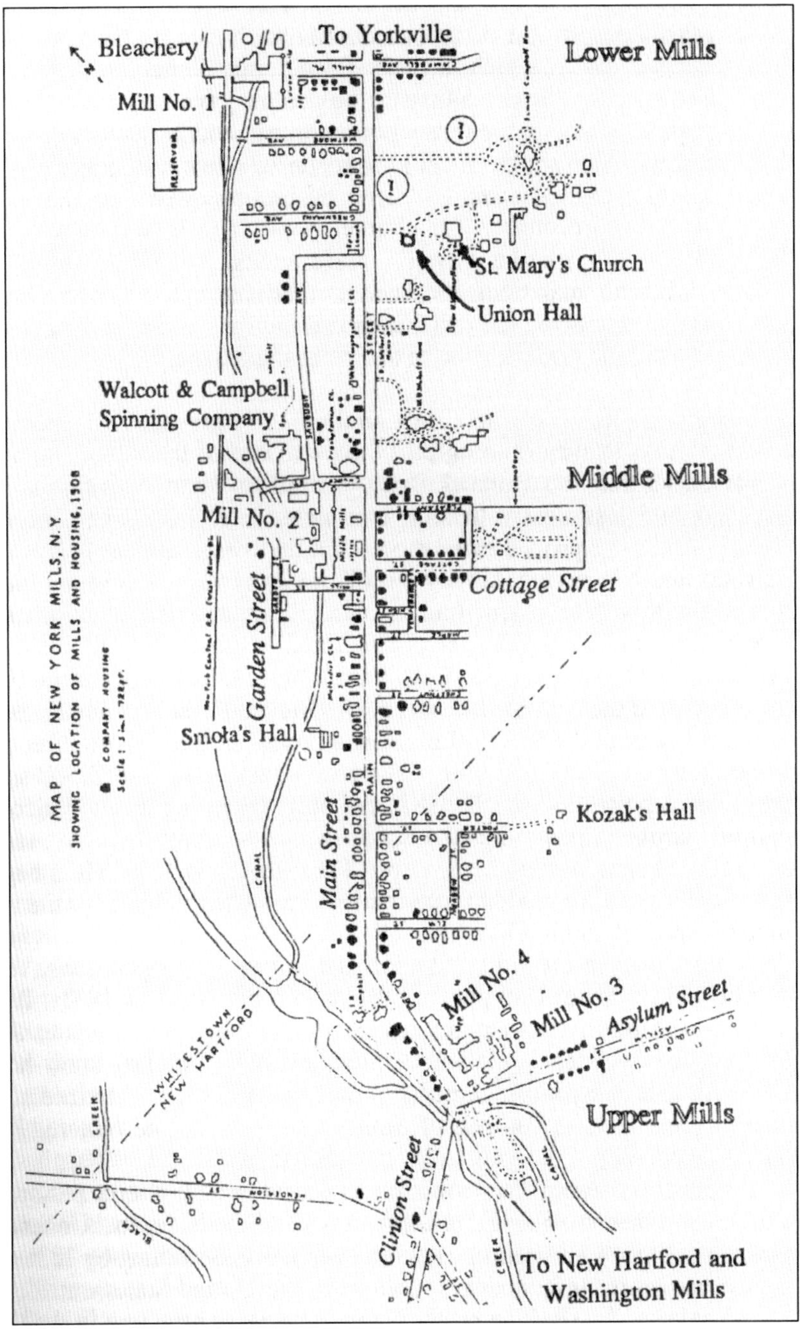

This map shows the village of New York Mills with prominent landmarks and the areas of Lower Mills, Middle Mills, and Upper Mills identified. (Authors' collection.)

ON THE COVER: The Polish settlers in New York Mills formed Local 753 of the United Textile Workers of America in 1911. It waged successful strikes against the A.D. Juilliard Company in 1912 and 1916. The 1916 cover photograph shows strikers in a tent encampment they constructed after being evicted from company-owned housing for going on strike. (New York Mills Historical Society.)

IMAGES
of America

NEW YORK MILLS

Eugene E. Dziedzic and James S. Pula

ARCADIA
PUBLISHING

ISBN 978-0-7385-9758-4

Published by Arcadia Publishing
Charleston, South Carolina

Printed in the United States of America

Library of Congress Control Number: 2012937938
For all general information, please contact Arcadia Publishing:
Telephone 843-853-2070
Fax 843-853-0044
E-mail sales@arcadiapublishing.com
For customer service and orders:
Toll-Free 1-888-313-2665

Visit us on the Internet at www.arcadiapublishing.com

Dedicated to the men and women of New York Mills who answered their country's call and especially to those who gave their lives in its service.

CONTENTS

ACKNOWLEDGMENTS

The authors extend their appreciation to all of those without whose assistance this book would have been impossible. In alphabetical order, they are: Stan Babiarz; Eugene Benman; Elaine Berdel; John Bialek; Jessica Bickford-Manson; Malio Cardarelli; Nancy Carey; Marie Dick; David Dudajek; Paul Dudajek; Nancy Dziedzic; Evelyn Edwards; Anne Flynn; Robert Glenn; Fred Griffiths; Gary Hadfield; Kathy Houghton; Brian Howard; Katie Kenney; Linda Kenney; Kim Kowal; Arthur Krawczenko; Dorothy Kula; Emil Labuz; Lou Ann and Willie Lachut; Bob Lalli; Rob Maciol; Heidi McManus; Walter Madej; Wieslawa Madej; Joe Marcinko; Wanda and Joe Miller; Elaine and John Montrose; Alfred Moses; Raymond Motyka; Eugene P. Nassar; Rose Nassif; Karen Neary; Vincent Nogas; Jackie, Jared, and Marty Pezdek; Al Piszcz; Wanda Potrzeba; Cheryl A. Pula; Fran Niziolek Pytko; Ron Pytko; Donald Rosinski; John Rumbutis; Carl Saporito; Gerald Showalter; Victoria Sobkowicz; Jim Stanek; Tom Stockton; Ed Synakowski; Ronald Szczepanski; Joan Kozak Unczur; Stan Walewski; Carol Wallace; Mildred Webb; Gene Wegrzyn; Joe (Wilczynski) Wilski; Donald Wisnoski; Davis Yohe; and Steve Zurek. The following organizations were also crucial to our research: New York Mills Historical Society, New York Mills Public Library, New York Mills Public School, the Oneida County Historical Society, and the Utica College Archives.

Sources of photographs are acknowledged in the captions with the following abbreviations: Babiarz-Pytko Oriskany Street Garage Collection (Babiarz-Pytko), Emil Labuz (Labuz), Wanda and Joe Miller (Miller), Elaine and John Montrose (Montrose), New York Mills Historical Society (NYMHS), New York Mills Public School System (NYMPS), Oneida County Historical Society (OCHS), Cheryl A. Pula (C. Pula), Fran and Ron Pytko (Pytko), Sacred Heart of Jesus Holy Cross Parish (PNCC), the State Historical Society of North Dakota (SHSND), Ed Synakowski (Synakowski), Utica College Archives (UC), Utica Public Library (UPL), Mildred Webb (Webb), Donald Wisnoski (Wisnoski), and authors Eugene Dziedzic and James Pula (Authors' collection).

INTRODUCTION

Shortly after the American Revolution, as the westward movement of settlers began to populate central New York, a sawmill and gristmill became the first commercial manufacturing concern in Oneida County when it opened in 1788. From this humble beginning grew a major textile manufacturing concern that gained a national reputation and spawned a village that carried its name—New York Mills.

When Benjamin Walcott Sr. arrived from Rhode Island in 1800, he quickly organized Walcott & Company as a joint stock venture and opened the Oneida Factory along the swiftly running Sauquoit Creek, an excellent source of waterpower. This marked the initial transformation of a small domestic industry producing cheesecloth for local dairy farmers into the beginnings of a major textile industry. The new company grew steadily. In 1810, the New York State Legislature incorporated the growing business as the Oneida Manufacturing Society. It eventually purchased the original sawmill and gristmill, constructed additional facilities, and in 1827 built a large stone facility midway between the original Oneida Factory and the former sawmill and gristmill. Financed by Benjamin Marshall, a wealthy New York City merchant, the new facility was named the "New York Mill" after his place of residence. The entire company was reorganized into the New York Mills Corporation in 1851. Its remarkable period of growth continued through the rest of the 19th century. In 1828, the mills produced 312,000 yards of shirting. By 1888, they were producing seven million yards of cloth annually. Ernest J. Savoie, in his history of the New York Mills Corporation, concluded that the "New York Mills goods were perennial favorites at national fairs and exhibitions where they usually claimed several of the top honors." One newspaper even complained that, "The New York Mills also exhibit their Shirtings and Cottonades. We think it hardly fair to allow these goods in exhibition, as they are sure to carry off the premium, and, therefore, men of less experience stand small chance."

As the industry increased in extent and importance, so did the community that grew up in its shadow. According to Rev. Charles B. Austin, in the early years the village contained only five farmhouses. Early employees commuted from nearby settlements. That soon changed. By 1830, two boardinghouses and 22 company-owned cottages provided lodging for workers, while a store, a church, and a school tended to their needs. From these humble beginnings, the village grew to a relatively constant population of between 3,000 and 3,500 people by the end of the century. A study of the community conducted by the New York Folklore Society about a century and a half later concluded that:

> The village's appearance soon attested to a sense of order and stability, which the company fostered in its dual attempt to draw workers and promote public favor. Both workers and proprietors lived near the mills: the workers clustered in company housing; the owners lived in finer homes, set somewhat apart by expansive lawns, but still close at hand. The workers' dwellings were brick or frame row houses and duplexes of simple but sturdy construction.

Frequently the cottages were built on a four-room New England plan, with sheds added in saltbox style. Churches and proprietors' homes displayed a full range of Victorian fashions, and the mills were kept up to date with the addition of stylish details such as cupolas and mansard roofs, which further reflected their owners' tastes. Such treatment of industrial buildings gave voice to a popular nineteenth-century philanthropic notion that home and work atmospheres could be rendered one in a company town, provided tasteful design be incorporated into the work environment. Of course, popular taste was synonymous with the owner's preferences, and while the mills and mansions received embellishment, workers' homes displayed only an occasional fan light over the door or a simple brick and stone design in the facade.

While the community grew in size, the energizing religious fervor of Rev. Charles Grandison Finney's Second Great Awakening preached that it was no longer sufficient for Christians to lead good lives. They must earn their way to salvation through good deeds. The argument convinced many to become active in the reform movements sweeping the northeast, feeding the growth of the temperance and antislavery reform movements in the village. New York Mills became known especially as a strident opponent of slavery, sending petitions to Congress and participating in organized opposition to the institution of human bondage. As early as 1842, the village's Presbyterian congregation adopted seven resolutions "strongly denouncing slavery," which Reverend Austin characterized as "forming as decided a document as the most radical might ask, and the last resolution urging that at the next meeting of the Presbytery it be requested to take measures to combat the great national evil." At least two ministers were dismissed when their commitment to the antislavery crusade proved less than their congregation expected. Nor were words the only weapon. When the Civil War erupted, the Walcotts and Campbells supported the Union with their influence and financial resources. The small village sent off some 150 men to preserve the Union and eliminate slavery, 25 of whom gave their lives for the causes.

Following the end of the Civil War, the nation experienced widespread industrial growth, as did New York Mills. During these years, the village operated as a classic company town, with the owners directly involved in the lives of the workers. Although the Walcotts and Campbells were generally considered to be paternalistic, providing churches, schools, housing, and other necessary community infrastructure, they were also strict. Looking back upon this era in 1915, a Utica newspaper explained:

> For know ye that there were rules and regulations about New York Mills in the early days as rigid as the blue laws of Connecticut. Along with their ingenuity, their industry, their enterprise, and frugality, the early settlers along the Sauquoit brought with them from New England deep religious convictions. No family could find employ in the great stone mill without a commendation from the minister. No family could continue in the employ that disregarded public worship, that was addicted to alcoholic habits, that failed to restrain its children properly, that ran into debt, which did or did not do a number of things to conform with the strict ideas of the founders.

Curfew bells rang regularly to control the beginning and end of the work day, to sound the lunch period, and to announce the time when everyone was to be home preparing for a good night's sleep to be fresh for the next day's work. Yet, despite the controls, the community was a generally harmonious place with only two minor worker protests in the 1890s.

By the 1880s, steam power replaced waterpower as the major source of energy. As in earlier years, most of the employees continued to be women and children, many of the latter being 14 years of age or younger. In the late 19th century, the company began attracting French Canadian workers to supplement the available labor supply, and by the first decade of the 20th century, it was seeking Polish workers from New England and Europe. The latter group would, over the years, grow to be the dominant force in the village, but upon their initial arrival, the Poles were merely the latest newcomers on the bottom rung of the socioeconomic order. By 1905, some 500 already resided in the village, and within 10 years, they had established a church, religious and secular organizations, a Union Hall, and a small but growing group of entrepreneurs.

In 1906, a major change in the life of the community came with the sale of the textile mills to the A.D. Juilliard Company of New York City, which abandoned the relative paternalism of the original owners in favor of maximizing profits. When this led to deteriorating treatment of the workers and reductions in pay, the Polish workers formed a union in 1911 and went on strike the following year. The strike lasted 27 days, idling 2,028 workers and in the end gaining wage increases and promises of better treatment. Yet, when ill treatment continued, workers voted to strike again in 1916. The company retaliated by evicting strikers from company housing, hiring private guards to rough up picketers, and importing strikebreakers. In a particularly bitter—but largely peaceful—strike that lasted 14 weeks and one day, the workers gained a 10-percent raise, reinstatement of all workers to their former positions, elimination of discrimination within the factories, and the reoccupation of company housing by those who had been evicted without the payment of rents for the period of eviction. The company also agreed to recognize Local 753 of the United Textile Workers of America as the workers' collective bargaining agent.

From the labor front, New York Mills next contributed to the battlefront, sending its sons off to fight in "The War to End Wars." At least 168 village residents served in the American armed forces, while others enlisted in the Canadian and Polish armies, the latter formed in Canada to serve with the Allied armies in France. Fundraising events were held, special collections were taken in the churches and at dances and concerts, and the New York Mills Corporation instituted a payroll deduction system for those who wished to purchase government bonds or contribute to the War Chest drive. Liberty Bond drives and Decoration Day celebrations in 1918 brought huge community support. Another significant contribution was made by Frederick C. Walcott. Before American entry into the war, Walcott became intensely interested in the fate of civilians caught up in the maelstrom of the European War. To assist in providing relief, he traveled behind the German lines into Belgium and Poland from 1915 to 1916 to direct relief efforts to those beleaguered peoples and was also concerned with Armenian, Serbian, and Syrian relief efforts. Visiting Poland in January and February 1916, Walcott became an immediate spokesperson for that ravaged land. He worked with piano virtuoso and statesman Ignacy Jan Paderewski in his Polish relief efforts and later was a major figure in Herbert Hoover's postwar relief programs.

The 1920s brought renewed prosperity to the village as it did to most of America, but the buoyant optimism of the decade came to a sudden end with the stock market crash in 1929 and the economic poverty of the Depression. The mills reduced output, people were laid off, and the population declined for the first time in a century. Others sometimes found work at great distances from home. Once the New Deal programs began, many joined the Civilian Conservation Corps and Works Progress Administration, taking jobs often far away but sending money home regularly to support their families. Still others joined the armed forces. Of the latter, 13 were present to witness the beginning of America's next great trial with the bombing of the Hawaiian Islands by the Japanese on December 7, 1941.

Once again, the village rose to the challenge. People collected used tin cans, old tires, and other scarce items for recycling to assist the war effort. Victory Gardens appeared throughout the community, high school students raised funds to purchase bonds, and the textile mills promoted salary deductions as a means of purchasing bonds over time. Residents continually surpassed by wide margins the goals set for them in War Loan Drives. In 1945, the community surpassed every other one in Oneida County in the margin by which it exceeded its goals. More than 800 men and women eventually served in the armed forces. In September 1942, the Utica *Observer-Dispatch* ran an article titled "Center of Patriotism: Single Street Sends 20 to War" on the tremendous contribution of New York Mills to the war effort. A particular focus was Maple Street, a short block of four company houses and one private residence whose 10 families had already given 20 sons to the service. Two years later, it ran a similar commentary on the village's contribution to the war effort. When it was all over, Gen. Ogden J. Ross, speaking at a rally in National Park, stated, "In my opinion, the Village of New York Mills sent more people to war, per capita, than any other community in the United States."

The postwar years brought yet another challenge and another major transition. For almost 150 years, the development of the village, its culture, society, politics, economy, and every other major formative factor of its collective life were molded by the permeating influence of the textile mills. Then they closed. Although traumatic at first, residents soon found employment in surrounding industries, and in the process, the village transformed itself from a one-industry town to a modern, suburban community. The change brought with it all the trappings of the new identity: a Little League program, a Cub Scout Pack, Boy and Girl Scout Troops, a Lions Club Youth International Youth Exchange Program, an Optimist Club, and new, community-wide celebrations. It also brought new construction to replace aging facilities: a school, library, firehouse, municipal building, and senior citizen residence.

From cottage industry to a national leader in textile manufacturing, from rural fields with an occasional farmhouse to classic company town, from a community oriented to a single major industry to a suburb with no major industry, New York Mills adapted to every change while retaining its unique historical identity.

One

An Industry Expands into a Village

In 1779, four New England investors purchased the Sadaquada Patent, a tract of land south of the Mohawk River and southwest of present-day Utica. In the years that followed, a cottage industry began producing cheesecloth to wrap the area's dairy products for market. In 1808, another group erected the first textile mill in New York State under a joint partnership named Walcott & Company. From a single, small wooden structure, the business grew steadily, adding factories until annual production rose to 312,000 yards of shirting by 1828, then skyrocketed to 1.8 million yards in 1831. The following year, *Lamb's Sectional Histories of New York State* noted that the mills "had the reputation of producing cotton cloth superior in quality to any other made in the United States." As production and reputation grew, the company began shipping most of its products to New York City for sale on the national and world markets.

As the company grew, it began to construct housing for its increasing workforce, building two-family cottages and boardinghouses first. The beginnings of a village soon attracted peddlers who later opened stores that were more permanent. The company began a rudimentary school, and its owners provided financial support for churches to tend to the needs of the workers. By 1851, the community numbered 2,000 people with Methodist and Presbyterian churches. Greatly influenced by the teachings of Rev. Charles G. Finney and his brand of religious activism, social movements such as temperance and abolitionism flourished among the residents. A Temperance Society opened a temperance saloon with nonalcoholic beverages, petitions were sent to Congress calling for the abolition of slavery, and ministers who were not sufficiently attuned to antislavery activism were replaced by those who were.

By the eve of the Civil War, one observer noted that, "Nowhere in the United States can there be found a factory village, which excels the New York Mills in the beauty of its natural surroundings, the neatness and comfort of the operatives' homes, the excellence of its schools, the prosperity of its churches, or the contented intelligence of its people."

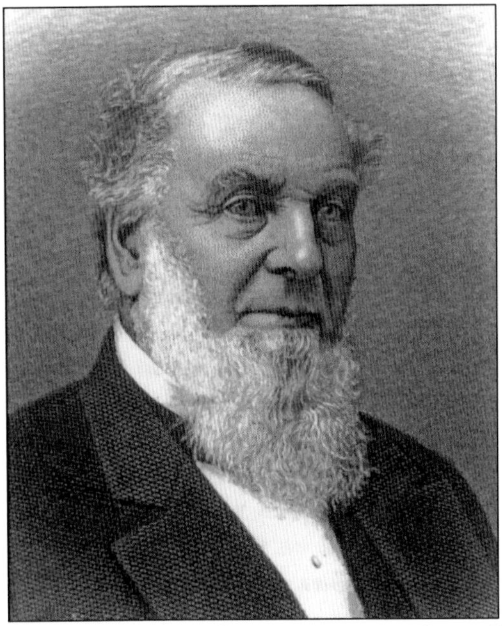

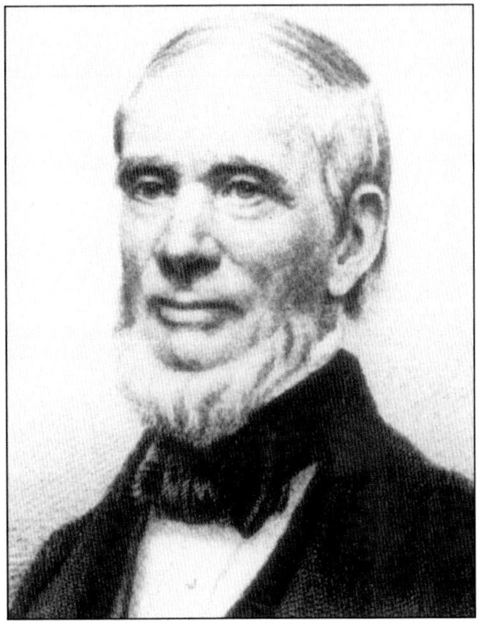

Born in Rhode Island in 1786, Benjamin Stuart Walcott Jr. (right) was brought to central New York by his father to become the superintendent of the first cotton mill established in the area. Samuel Campbell, who was born in Scotland in 1809, arrived in 1831 with mechanical knowledge that turned the factory into the largest producer of cloth outside New England by 1838. (OCHS.)

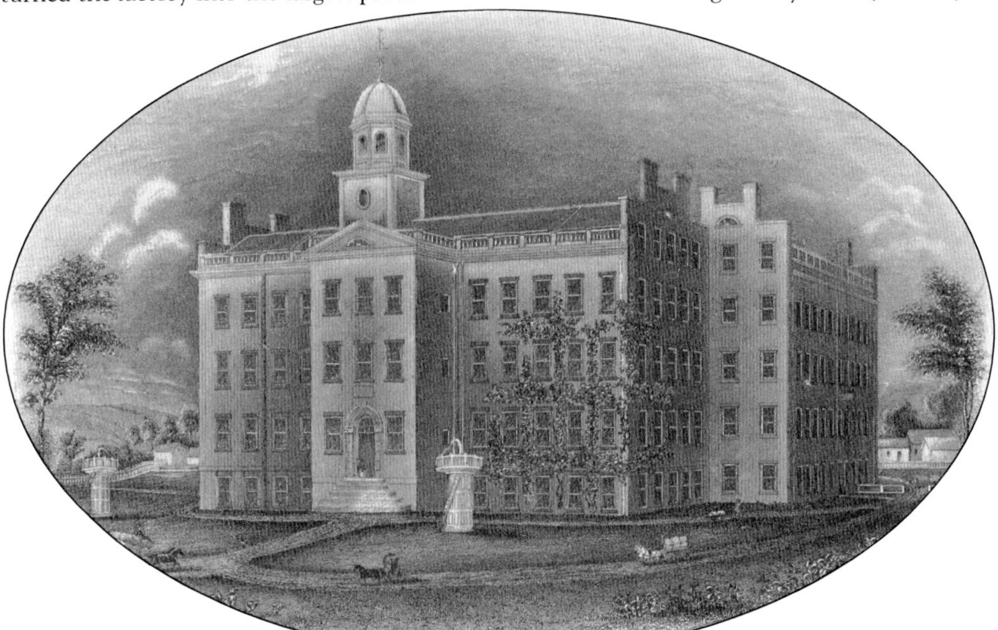

The Oneida Manufacturing Society built the first cotton mill west of Rhode Island on three acres and 61 rods of land it purchased in the Sadaquada Patent for $538. The mill was operated by Walcott & Company, a joint stock company including investors Benjamin Walcott Sr., Benjamin Walcott Jr., Seth Capron, Asher Wetmore, Newton Mann, William Cheever, Thomas Gold, and Theodore Sill. (OCHS.)

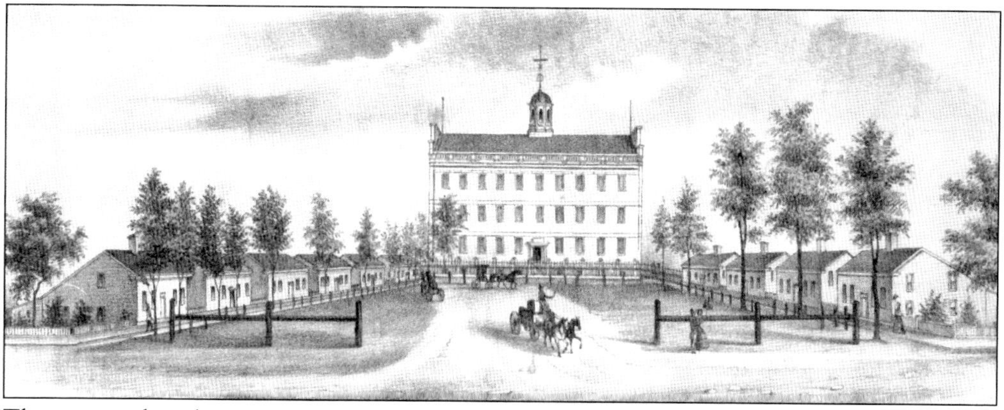

The cotton that the Oneida Factory used was first sent out to local families to be picked, or prepared for carding, prior to the invention of the picking machinery. It employed approximately 80 people to operate 384 spindles with the yarn it produced being sent to local homes to be woven into cloth about 27 inches wide. They received 5¢ to 6¢ per yard. (OCHS.)

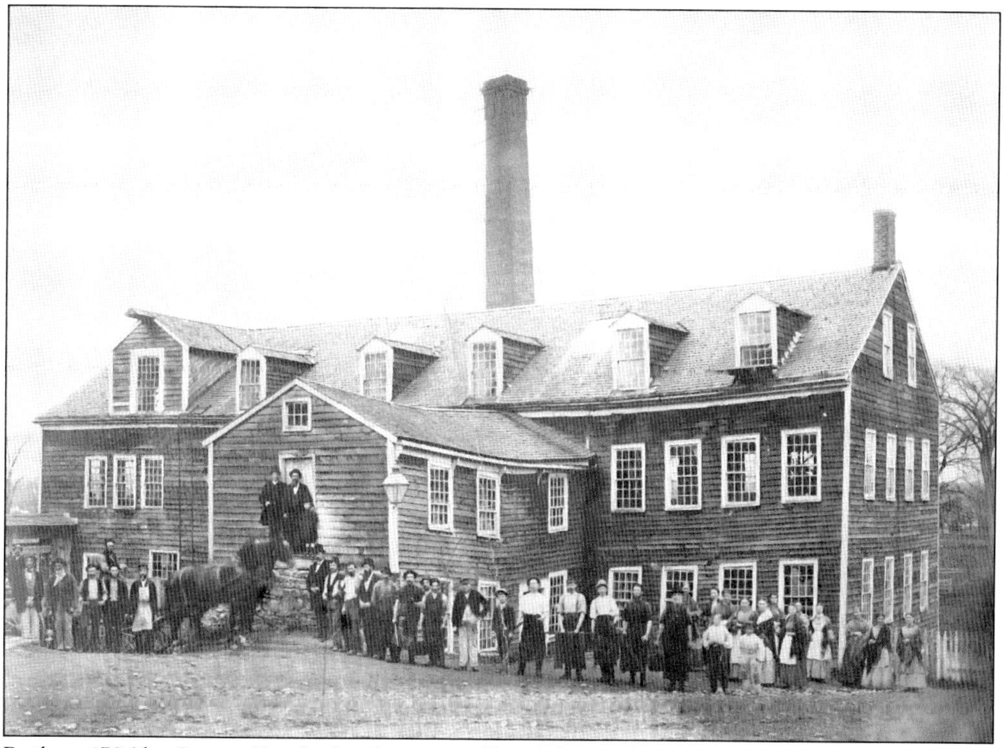

Built in 1796 by George Doolittle, this gristmill and flourmill was the only flour factory within 40 miles. Since it used French burr millstones, it was known as the Burr Stone Mill, a name it gave to its location on Burrstone Road. It was converted about 1820 to a cotton mill by Benjamin S. Walcott when he married Doolittle's daughter, and in 1835 the mill became a dye house. (OCHS, NYMHS.)

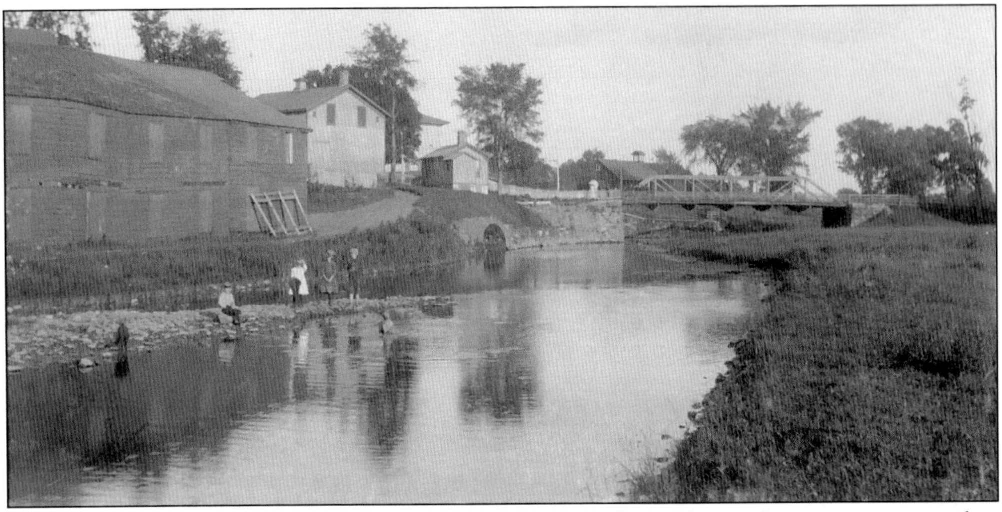

In the early 19th century, Sauquoit Creek was a relatively stable, swift-running stream that dropped some 1,014 feet over a 17-mile stretch. Walcott & Company found it an ideal source of the waterpower needed to fuel the new machinery it began to install. This dyke was installed for better water control and to funnel the stream into where its energy could be tapped. (NYMHS.)

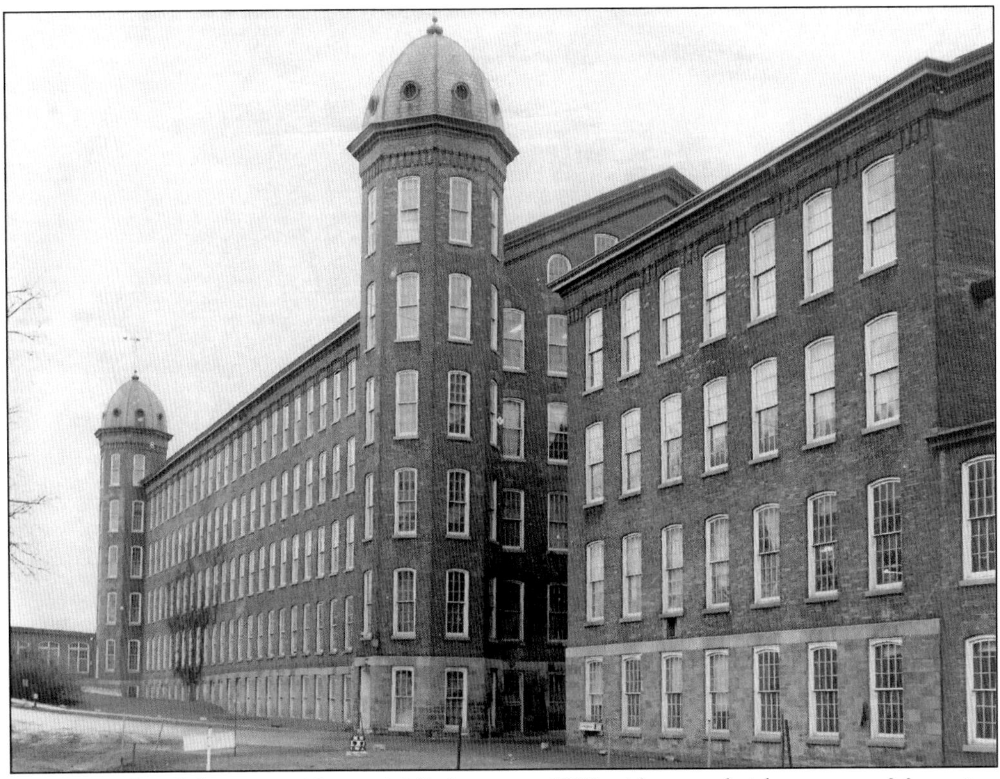

Walcott replaced the original Lower Mills factory in 1880 with a new brick structure. Measuring 235 feet by 78 feet, it contained weaving rooms in the basement and on the first floor and spinning equipment on floors two, three, and four. Breaking with the former Federalist architecture, this ushered in a freer style of construction thereafter. This new facility became known locally as the "No. 1." (NYMHS.)

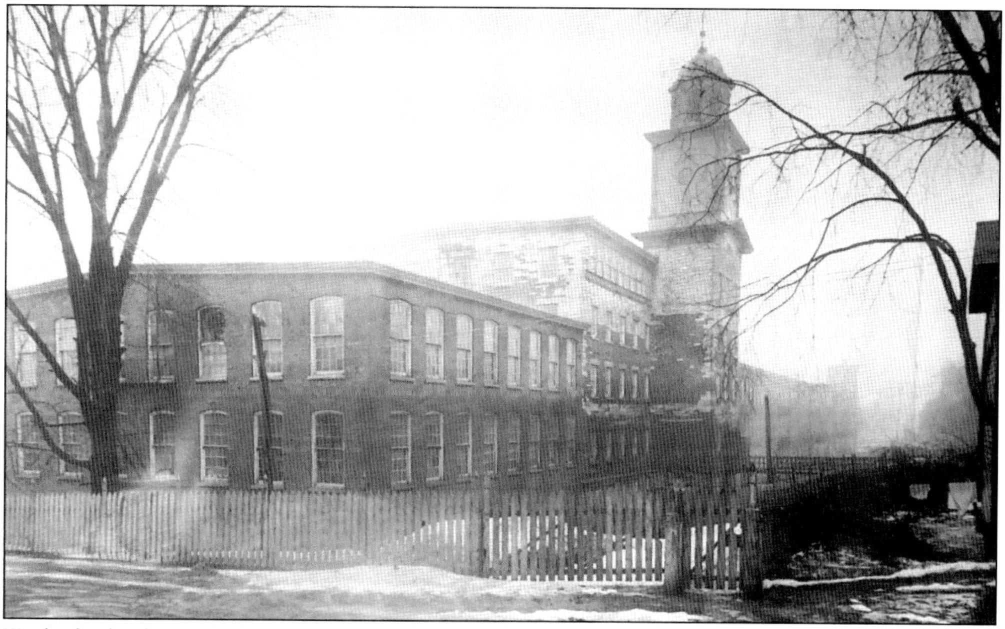

With the backing of wealthy New York City merchant Benjamin Marshall, Benjamin Walcott Jr. opened the New York Mills Company in 1827 on 89 acres of land. Constructed at a cost of $65,199.14, the five-story stone factory in the Federalist design contained 42 power looms capable of producing the finest-quality cotton sheeting and bringing the company national attention. This later became known as the "No. 2." (Authors' collection.)

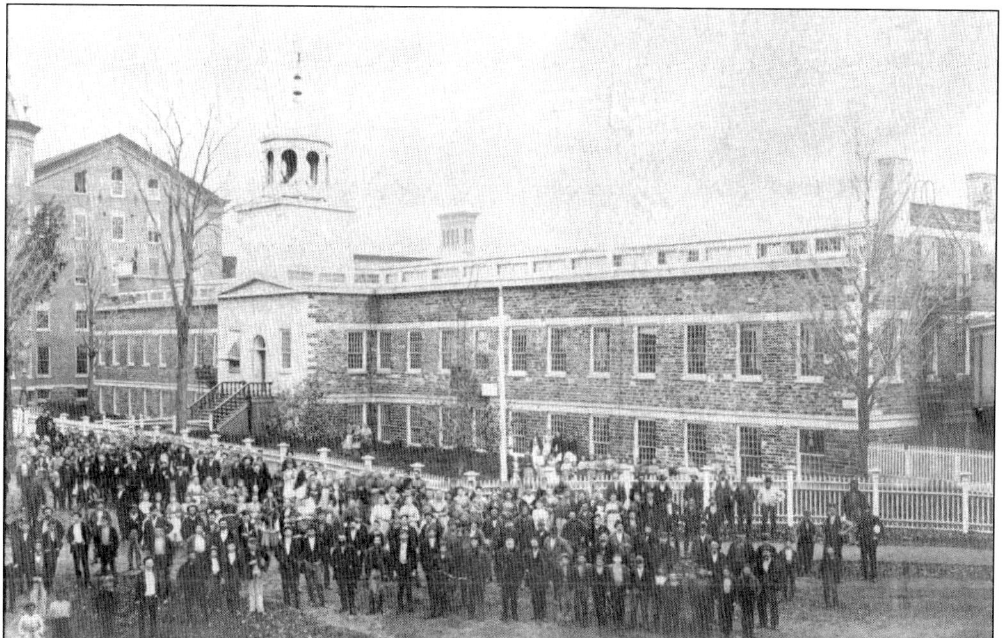

In 1839, Marshall and Walcott began construction on a two-and-a-half-story stone mill containing new Arkwright machinery. Located across the street from the original Burr Stone Mill, it was 229 feet long. When operational in 1840, it propelled the New York Mills into the top 10 largest producers of cotton yarn in the country. The new structure became known as the "No. 3." (OCHS, NYMHS.)

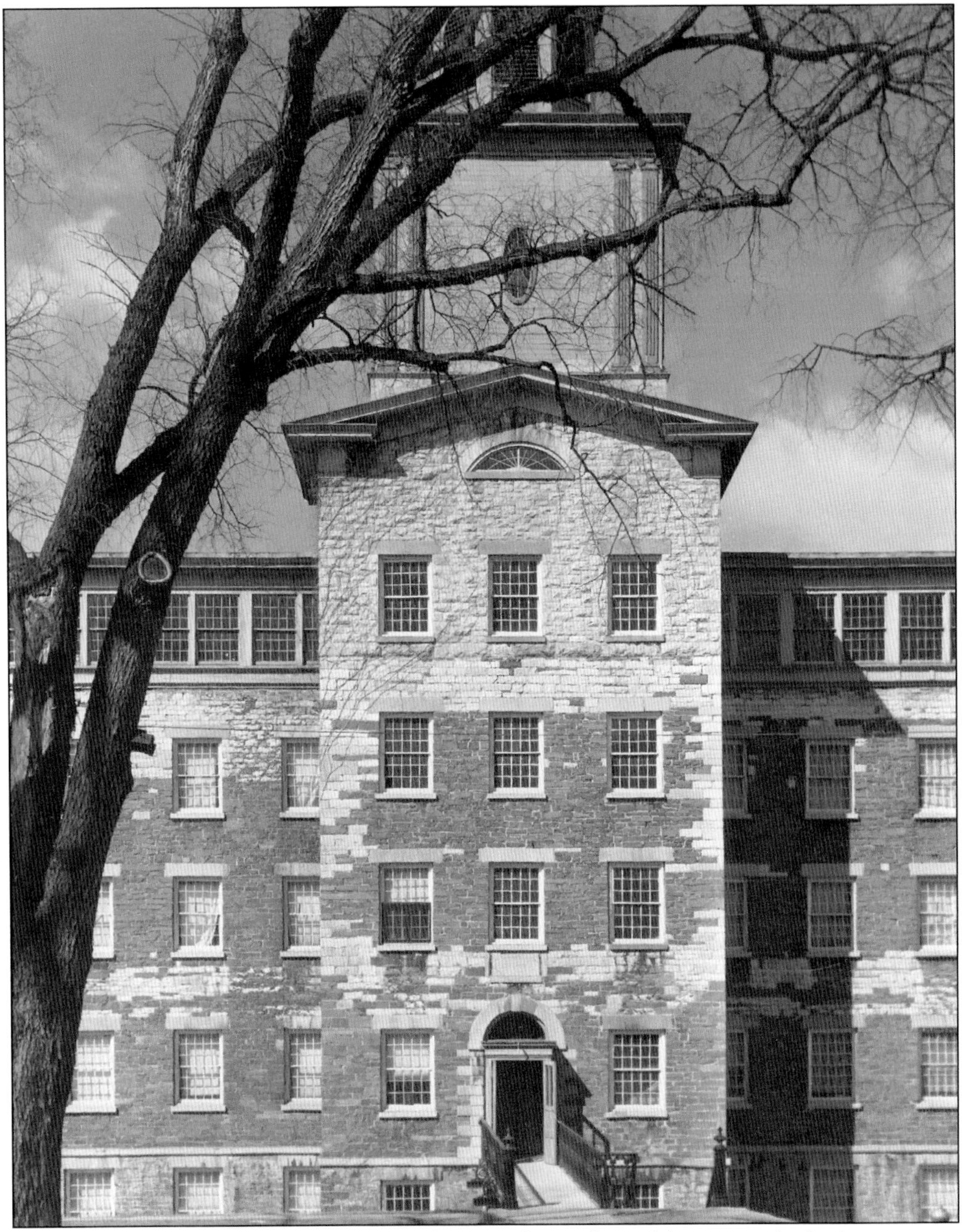

Completed in 1827, the No. 2 was 120 feet long, 20 feet wide, and five stories high. As soon as it was completed, a five-story wing was added, and in 1830, another addition was completed to house a machine shop and picking room. It was later expanded in 1852, and again in 1884 and 1904. At the time of its original construction, a canal was built to channel water from Sauquoit Creek to power the equipment, with the stone being quarried in Clinton and Little Falls. The original Federalist architecture pictured here featured clean rectangular lines capped by an octagonal cupola with balustrades along the edge of the roof. Federalist architecture formed the basis for much of the Walcott and Campbell business construction well into the 1850s and occasionally beyond, even manifesting itself in some of the private homes. (OCHS.)

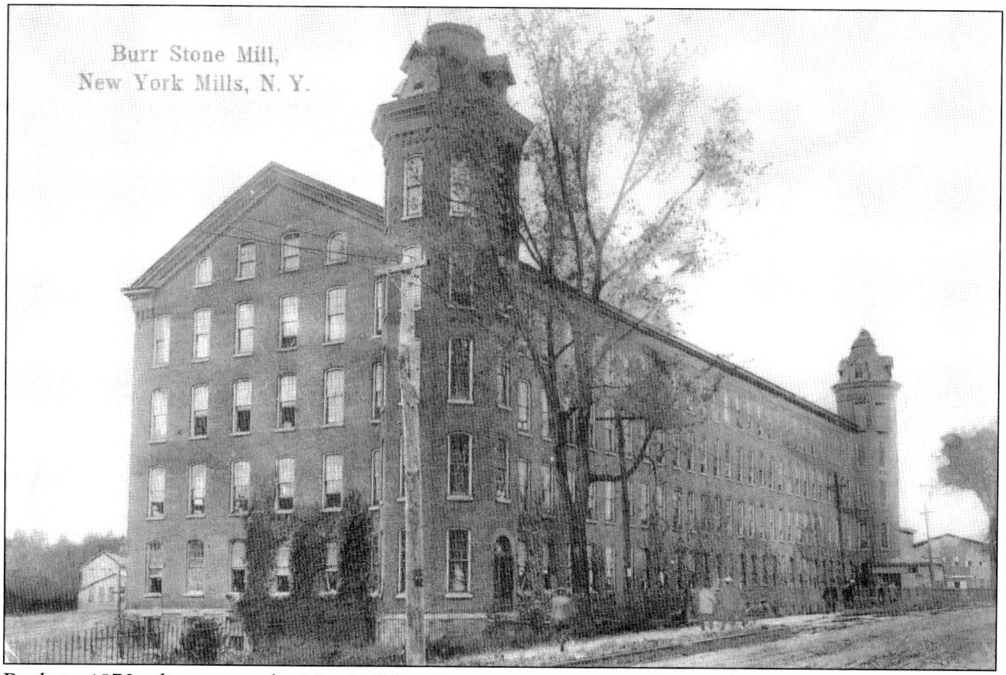

Built in 1870 adjacent to the No. 3, "No. 4" was a massive new four-and-a-half-story brick factory some 250 feet long and 70 feet wide. The first floor contained weaving machinery, while floors two, three, and four were dedicated to carding, spooling, warping, and spinning. It was distinguished by its two towers capped with octagonal Mansard roofs. (NYMHS.)

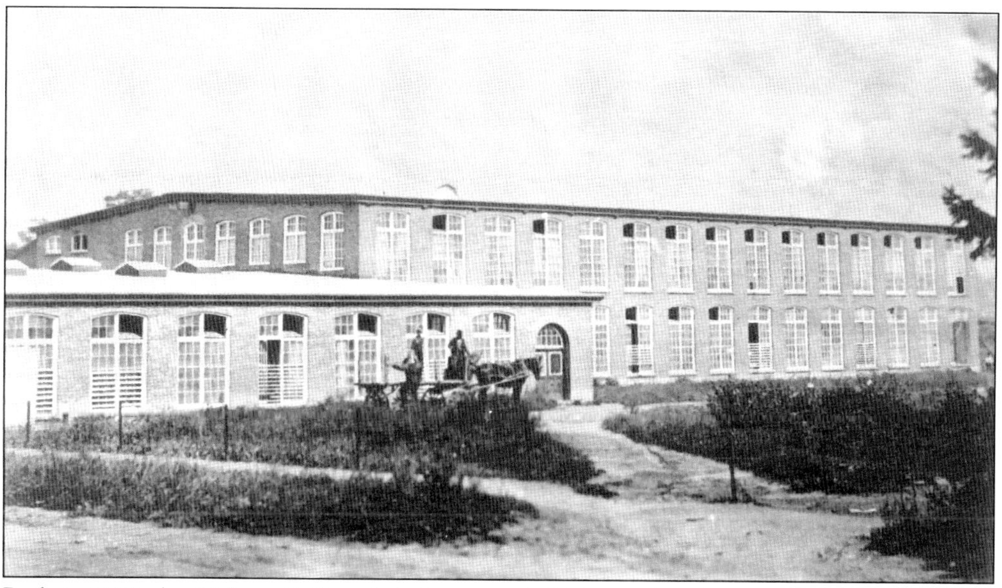

Built in 1902, the Walcott & Campbell Spinning Mill, located on Sauquoit Street, was initially owned and operated by the two men separate from the other mills. In the mid-20th century, the left section of the building housed the Nehi Bottling Works, while the section on the right was razed to construct a company manufacturing concrete pipe. (NYMHS.)

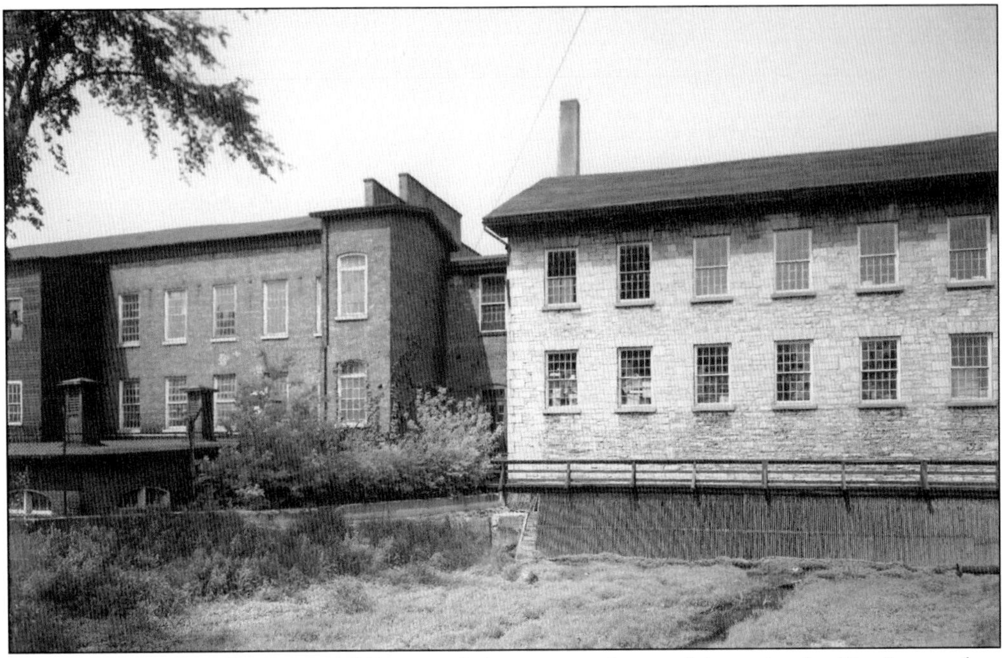

The wheelhouse in the Lower Mills provided power to the machinery in the No. 1 complex. To the left, just below the wheelhouse walls, can be seen part of the stone retaining wall that carried the waters of Sauquoit Creek through the mills to create the power that kept them running. (OCHS.)

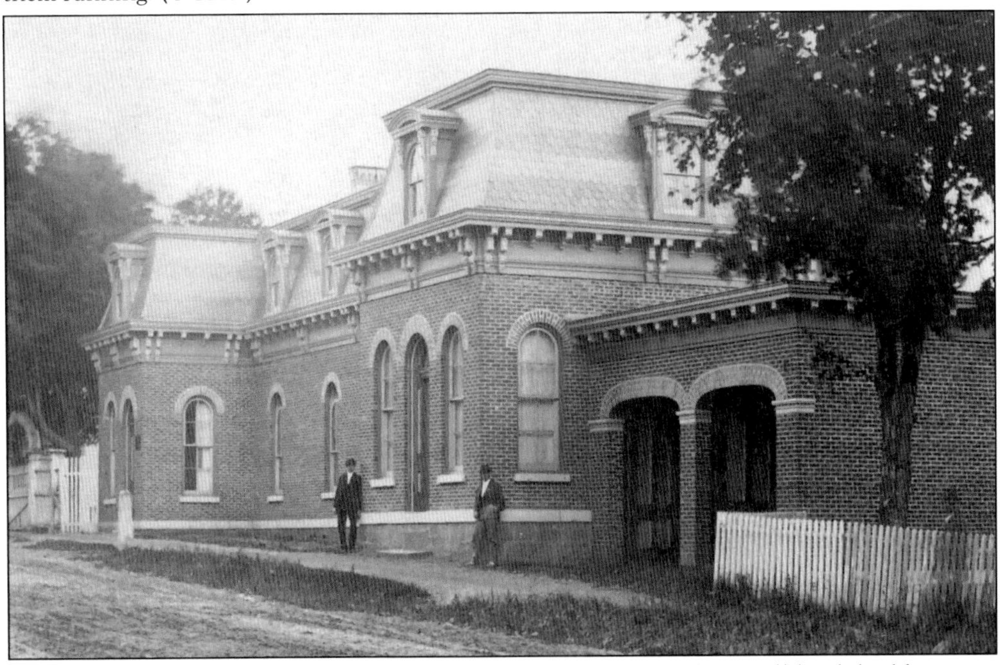

Central offices for the New York Mills factories were located in this small brick building near the center of the village on Main Street in front of the No. 2. This building included the senior management offices, central payroll office, purchasing, and other company-wide offices in a rather compact space. (NYMHS.)

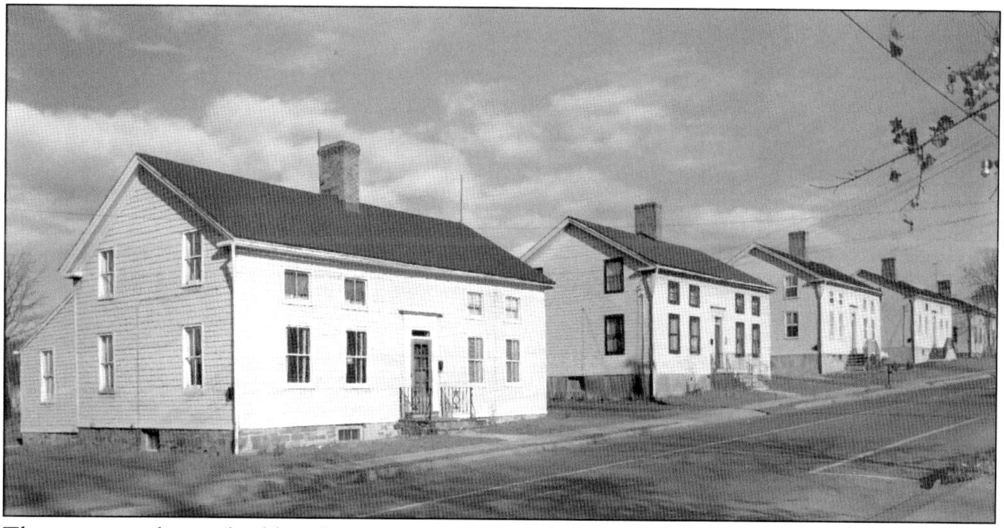

The company began building homes to rent to workers in 1825 with 22 structures on Garden and Cottage Streets and across Main Street from the No. 2 mill. Each 1.5-story frame structure accommodated two families. Later, rear lean-to additions were added. Of clapboard construction, they had shingle roofs and rubble foundations but were exceptionally durable as seen in this photograph taken some 150 years later. (Authors' collection.)

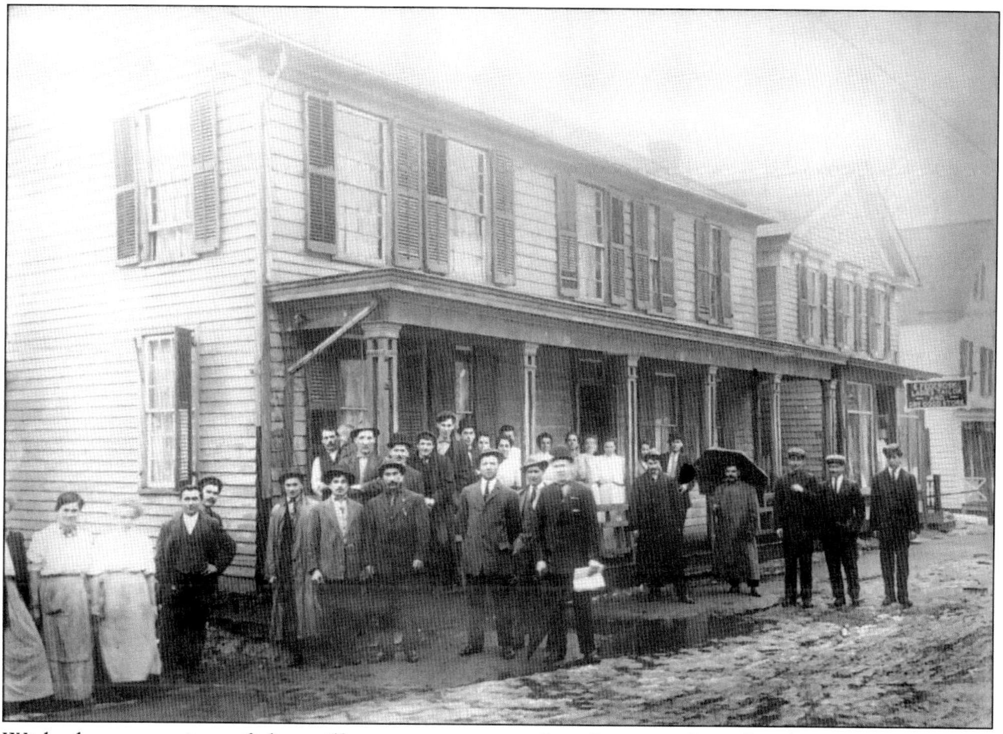

With the expansion of the mills, company-owned and private boardinghouses appeared to accommodate the increasing workforce. These were generally segregated by gender. This photograph depicts the Number Three Boarding House located at 500 Main Street in the Upper Mills. The building to the right is Stanley Federoski's Ladies Tailor and Dry Goods Store. (Authors' collection.)

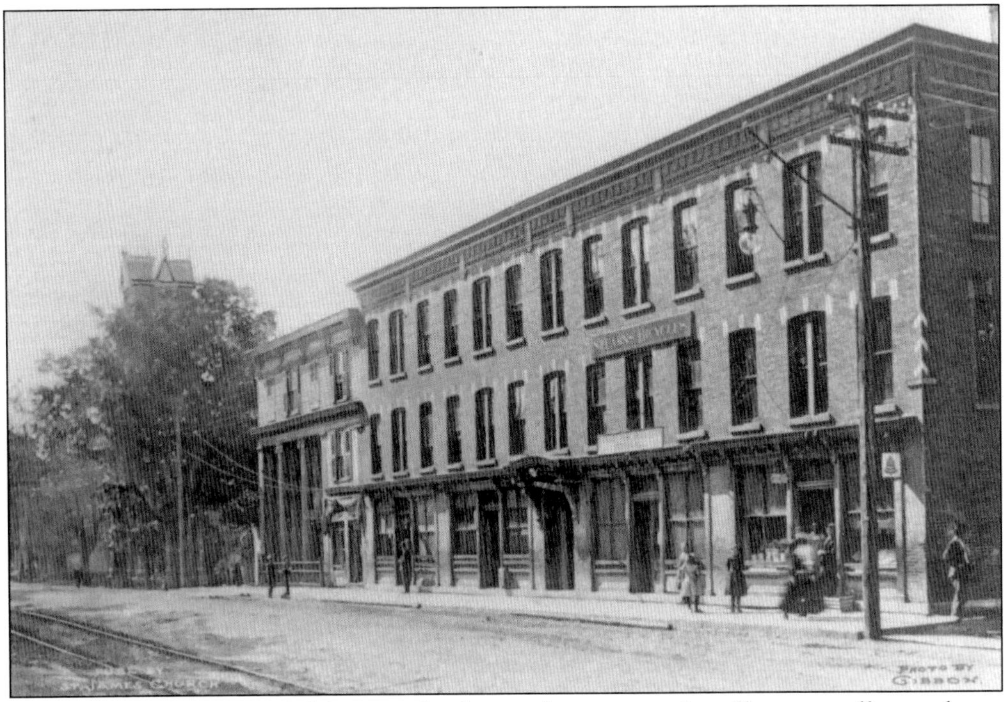

The Brick Block contained club rooms for the textile company, the village post office, and over time, a variety of small businesses including a grocery store and tavern. The prominent sign here advertises Stearns Bicycles. Located near the center of the village across Main Street from the fire department, it burned in September 1983 and was replaced with a parking lot. (NYMHS.)

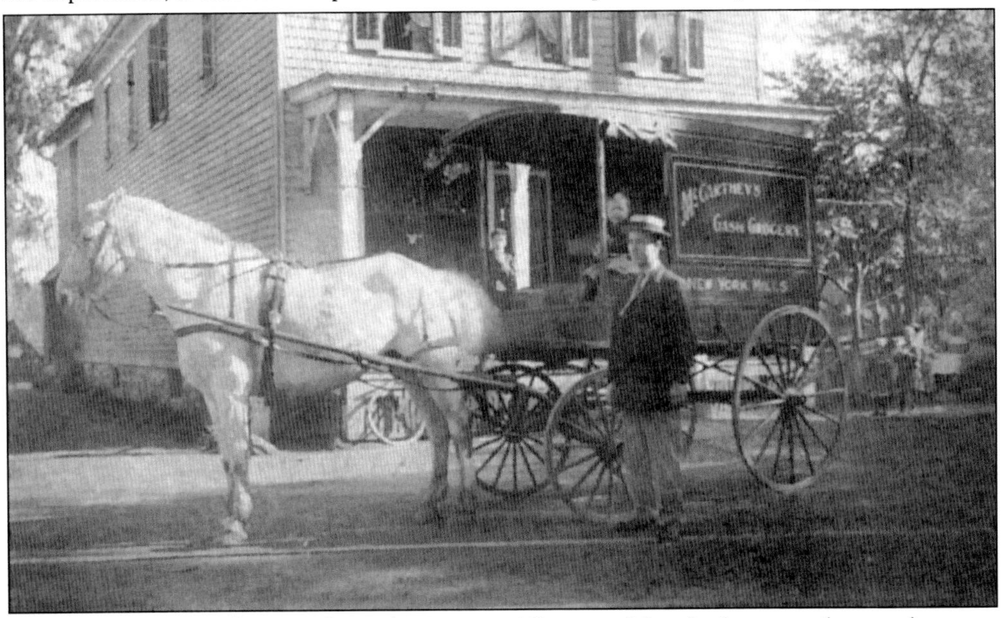

Many of the village's first merchants began as peddlers, or if they had a store, they used wagons to deliver merchandise and expand their reach to customers. William McCarthy operated his business out of the Upper Mills. The corner where New Hartford and Henderson Streets intersect Clinton Street was called McCarthy's Corner. (NYMHS.)

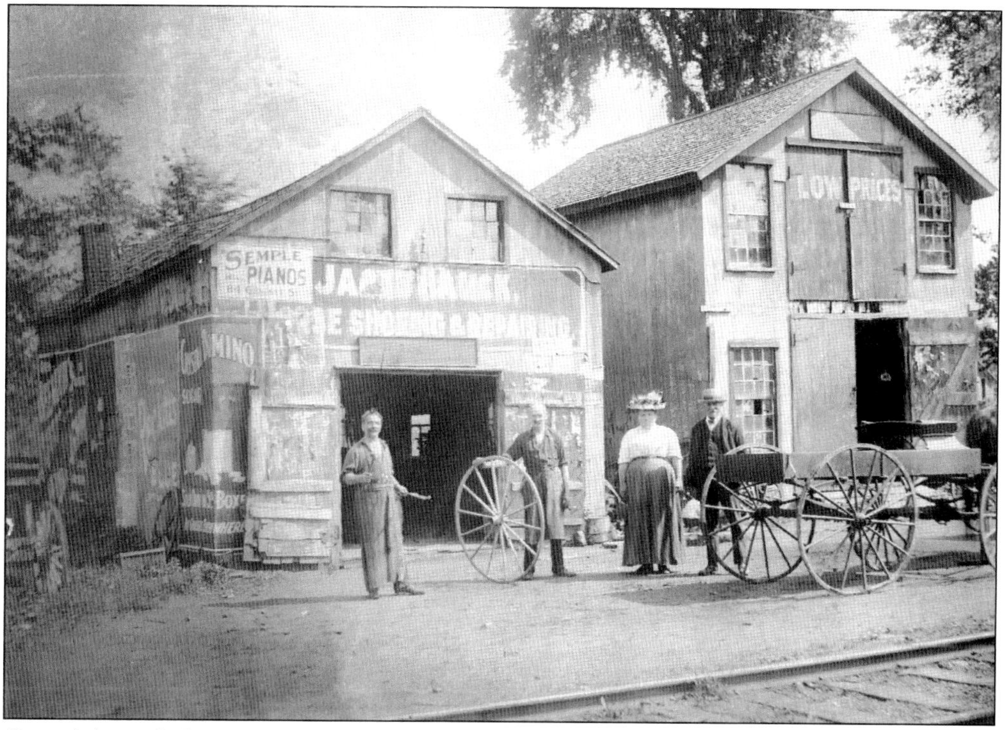

One of the early businesses was Jacob Hauck's repair shop on Clinton Street just south of the Sauquoit Creek bridge. Specializing in horseshoeing, wagon restoration, and other kinds of repair work, his entrepreneurial flair also appears to have included advertising. (NYMHS.)

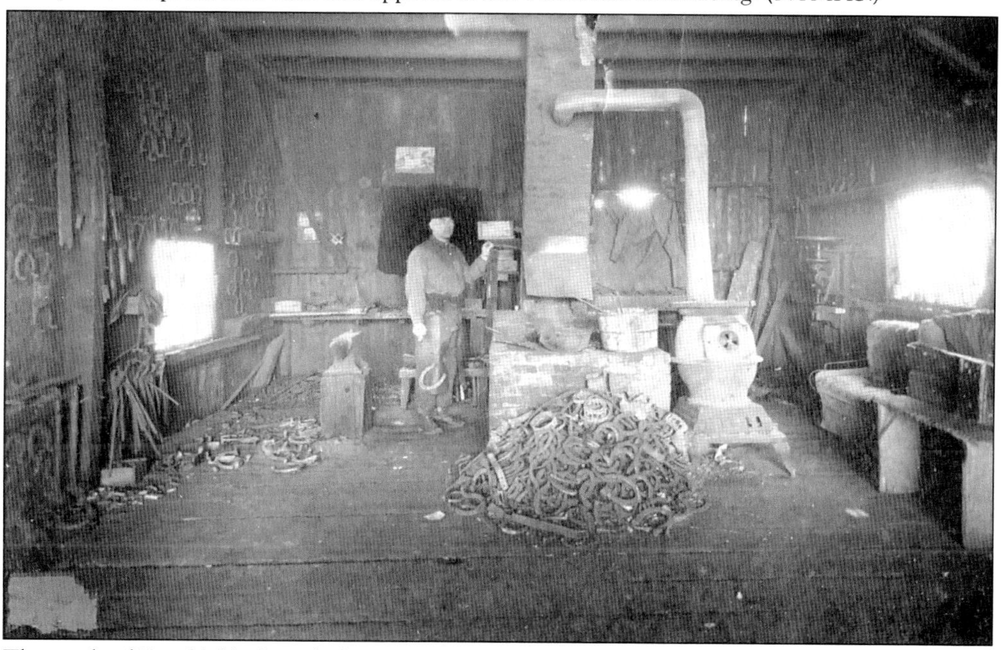

The inside of Hauck's blacksmith shop looks Spartan, but the number of horseshoes piled on the floor and affixed to the walls testifies to the major portion of his business. An anvil is clearly visible to the left, what appears to be a kiln in the middle, and a potbelly stove to the right. (NYMHS.)

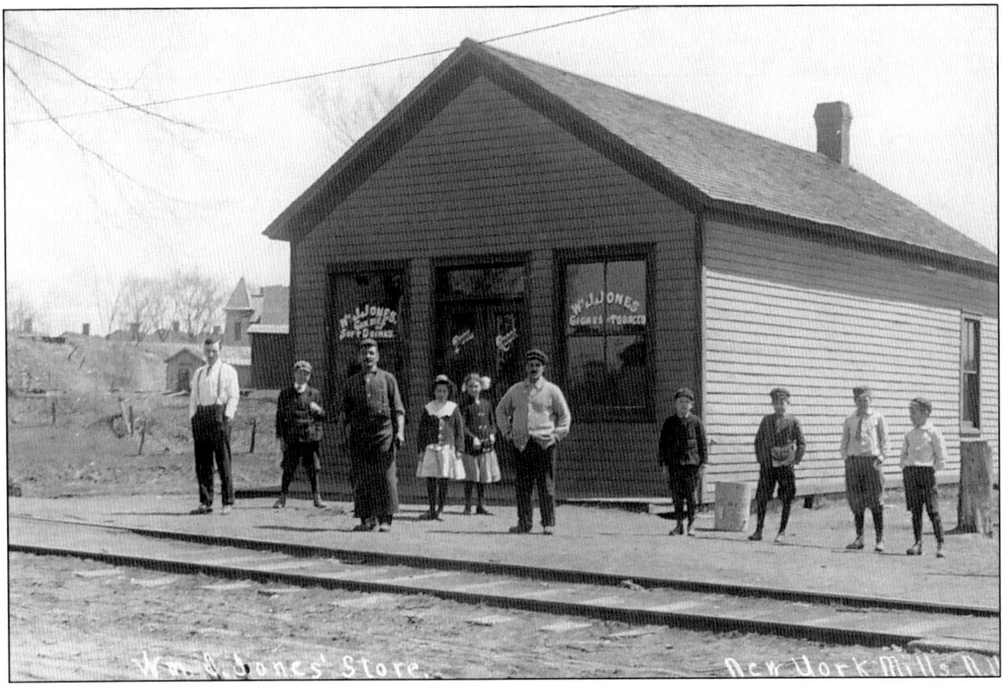

This general store, owned by William J. Jones, served the Upper Mills from its location on Clinton Street just south of the bridge over Sauquoit Creek. William Jones stands on the far left. Just over his head, the spire of the Upper Mills School is visible. (Pytko.)

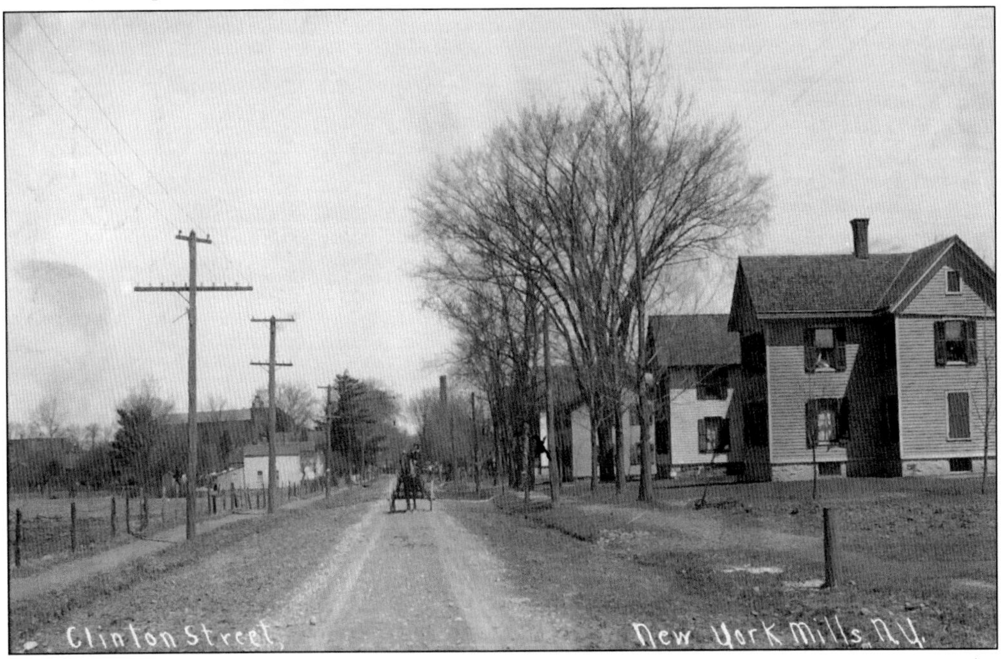

This photograph shows Clinton Street looking northeast toward the No. 3 in the 1890s. At the time, this was on the very edge of the village as can be seen from the open fields in the foreground and to the left. Note the horse and carriage approaching the photographer and the smokestack from the mills at the foot of Burrstone Road in the background. (NYMHS.)

The Methodist Church was organized in 1826 under Rev. Charles Giles with the strong support of Benjamin Walcott Jr. The church pictured here was the third Methodist church building. It was dedicated in June 1873. The brick building contained a basement for a Sunday school and an auditorium on the first floor. It has been described as being in the functional style of the period. (NYMHS.)

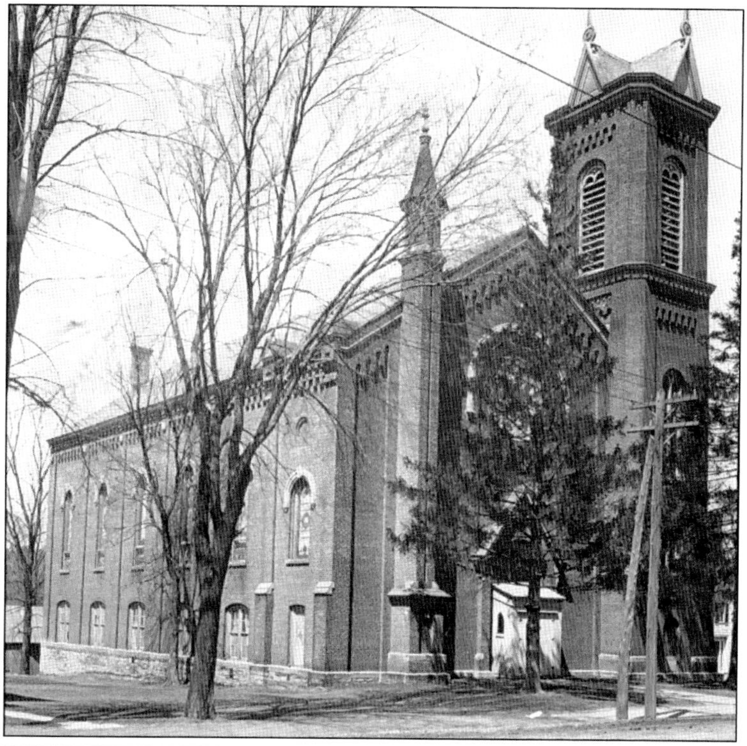

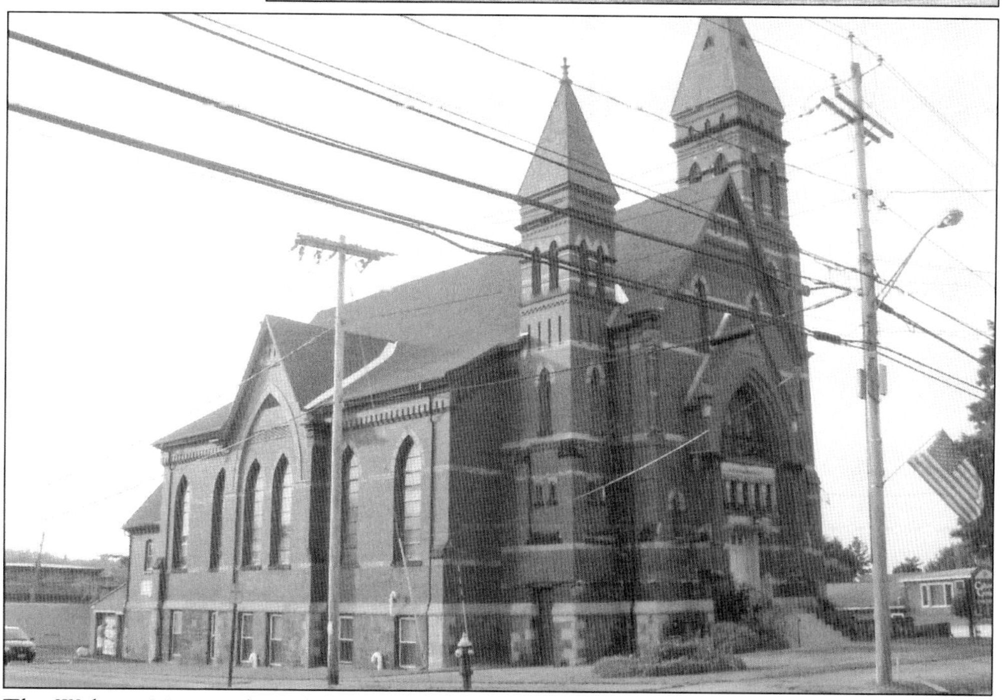

The Walcott Memorial Presbyterian Church formed in 1831 under Rev. George Foote. This building, dedicated on February 8, 1882, was designed by the Syracuse firm of H.N. White and F.J. Grodavent using Gothic arches and windows. The basement had a Sunday school, library, and kitchen, while the first floor contained an auditorium and gallery. (NYMHS.)

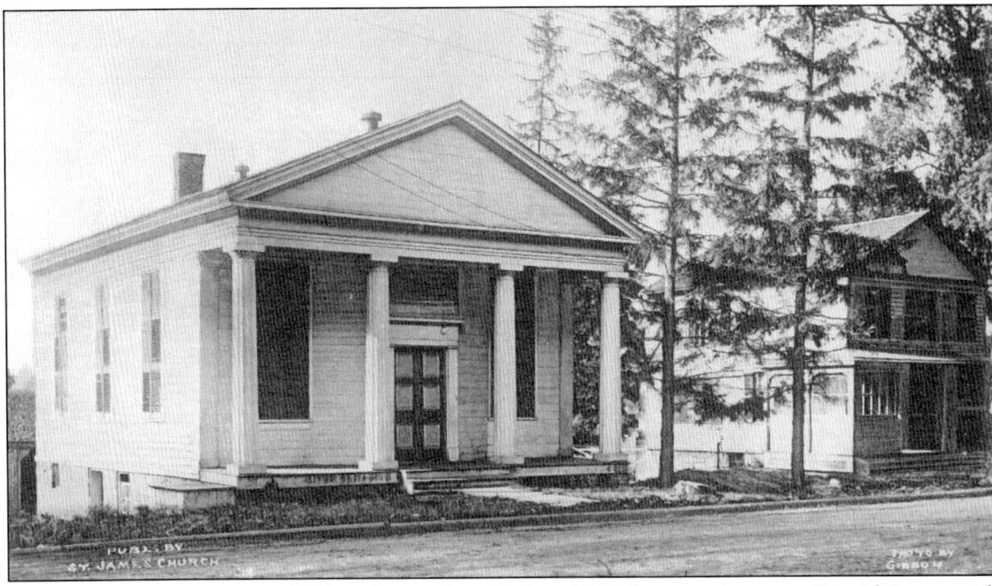

The Welsh Congregational Church was founded in 1847. Its president was David Jones with vice president Thomas Ellis, secretary Owen Baxter, and trustees Ellis Ellis, Owen Griffiths, David Jones, William Roberts, and Benjamin Walcott. In 1852, they erected a building on Main Street in the Greek Revival style with four Doric columns. It contained an auditorium and vestibule. (NYMHS.)

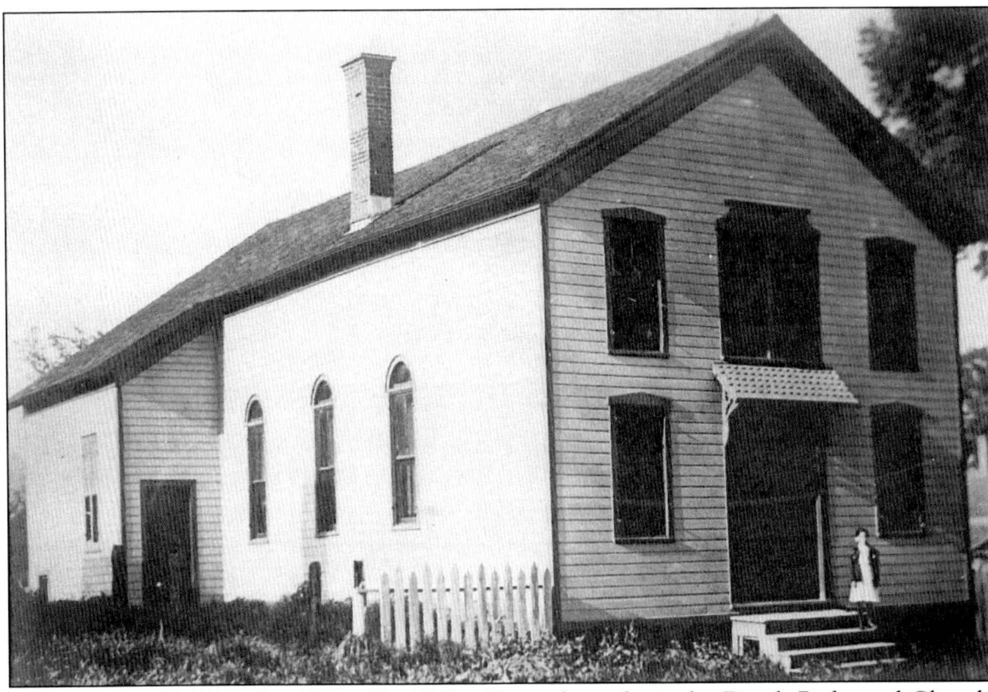

Although small in number, the village's Dutch residents formed a Dutch Reformed Church, known locally as the Holland Church. Its church was located on the corner of Cottage and Pine Streets, a location that later became a store and then private residences. At the time of the church's construction, Pine Street was known as Temperance Street. (NYMHS.)

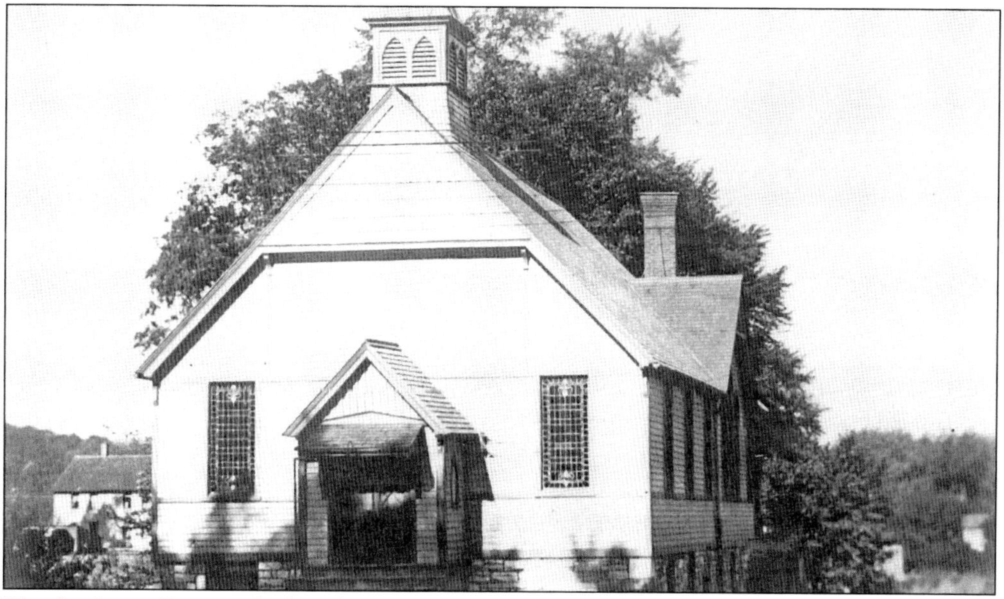

The St. James Episcopal Church, organized mainly by residents of English descent, opened a wooden church on the corner of Main and Walcott Streets in 1881. When the congregation dwindled, the building was eventually sold to a Polish group as home to the Sacred Heart of Jesus Polish National Catholic Parish in 1924. (NYMHS.)

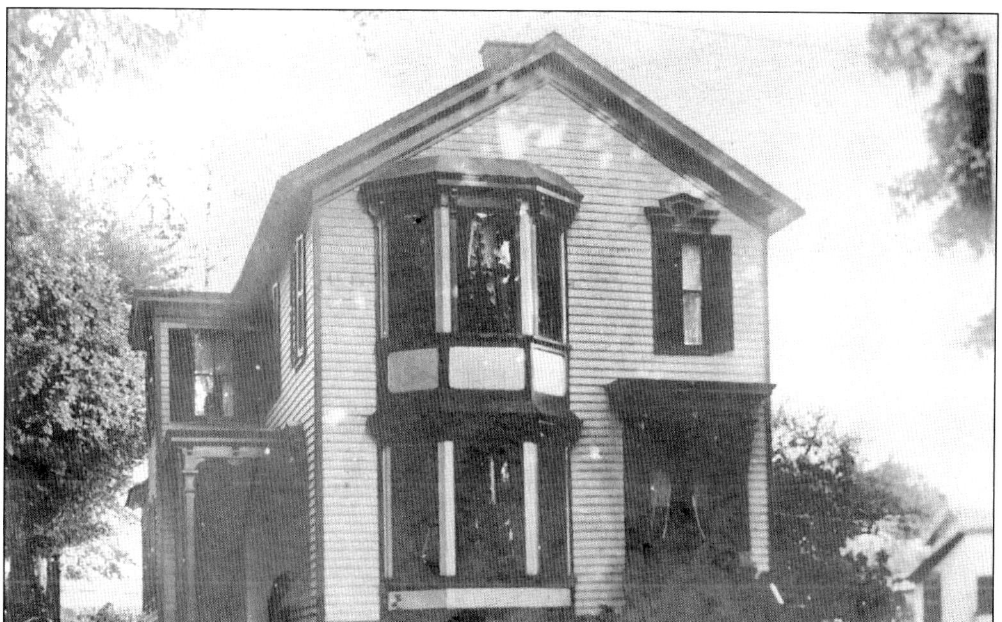

This two-story building on Main Street in the Middle Mills served as the parsonage for the Methodist Church. The residence hosted many visiting regional and national dignitaries since the church was very active in early social movements and later in the missionary movement to send people abroad to spread Christianity to Asia, Africa, and other areas around the world. (Pytko.)

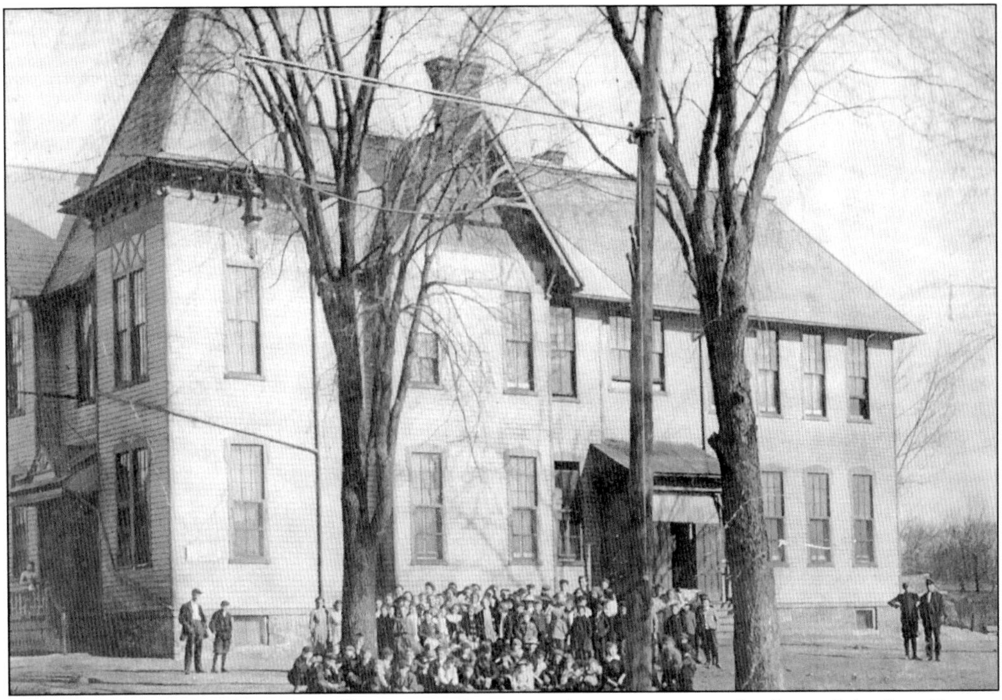

One of the first schools in the village was a large wooden structure at the foot of Asylum Street, which later became Burrstone Road. Children also used the grounds as a baseball field and general playground until the structure burned and was replaced by a dye house serving the textile mills. (NYMPL.)

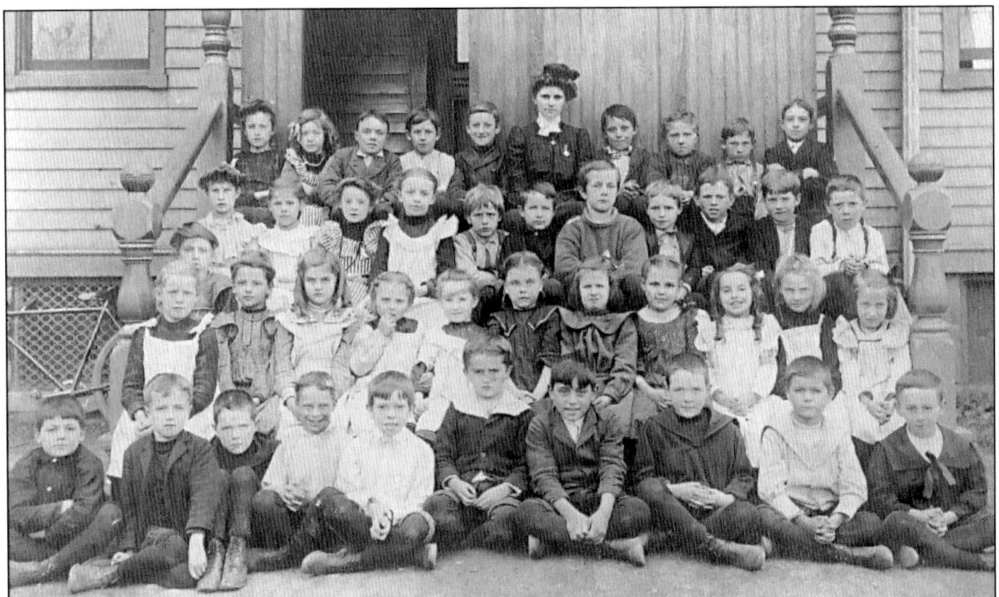

Sarah Roberts's students posed for this photograph at the Asylum Street school in 1890. There are 42 students in the photograph, which was relatively typical of the large class sizes that early teachers were assigned. Note also the students' apparel, which judging from other pictures of the time period, appears to have been standard school attire. (OCHS.)

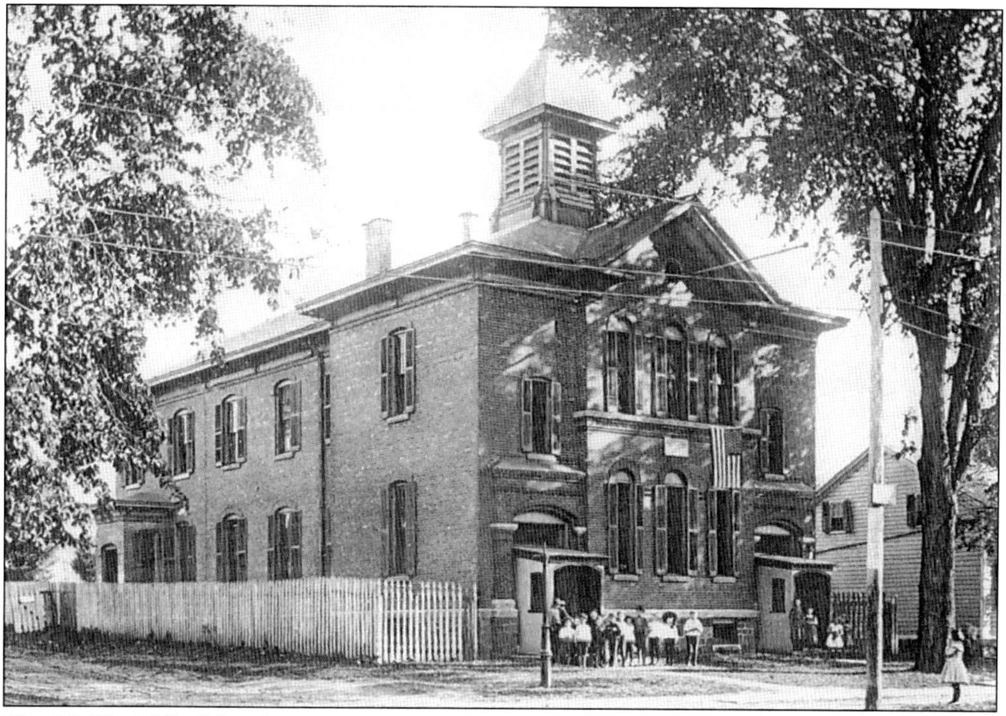

The Middle Mills School was a brick structure that also housed the small public library in the rear. Later, a one-story addition was constructed on the front to serve as a firehouse. Many residents fondly remember how a colorfully lit Christmas tree would be erected on top of the firehouse, which could be seen from a distance along Main Street. (NYMHS.)

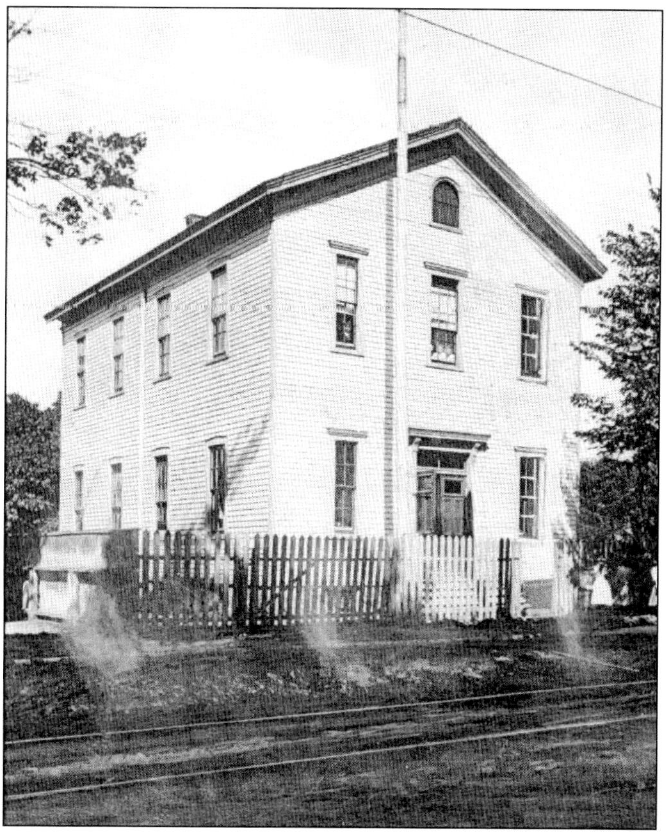

The Lower Mills School, located on Main Street close to the Yorkville village line, served the northern end of the community in the area of the No. 1 mill. Before the consolidation into a new brick schoolhouse in the Upper Mills, village children attended any of three smaller schools in the Lower, Middle, and Upper Mills. (NYMHS.)

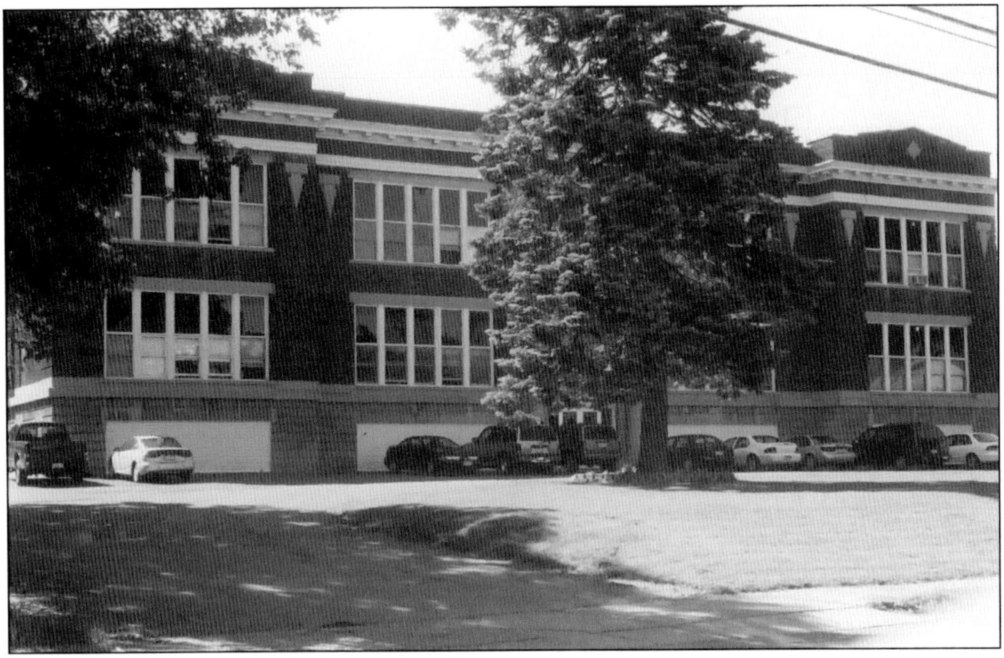

In 1916, a brick school for high school and elementary classes was constructed on Main Street at a cost of $80,000. With a capacity of 385 students, the building housed nine classrooms, a library, a laboratory, three recitation rooms, a principal's office, and a combined assembly hall and gymnasium that could seat up to 450 people. This image shows the structure in its later configuration as an apartment complex. (Authors' collection.)

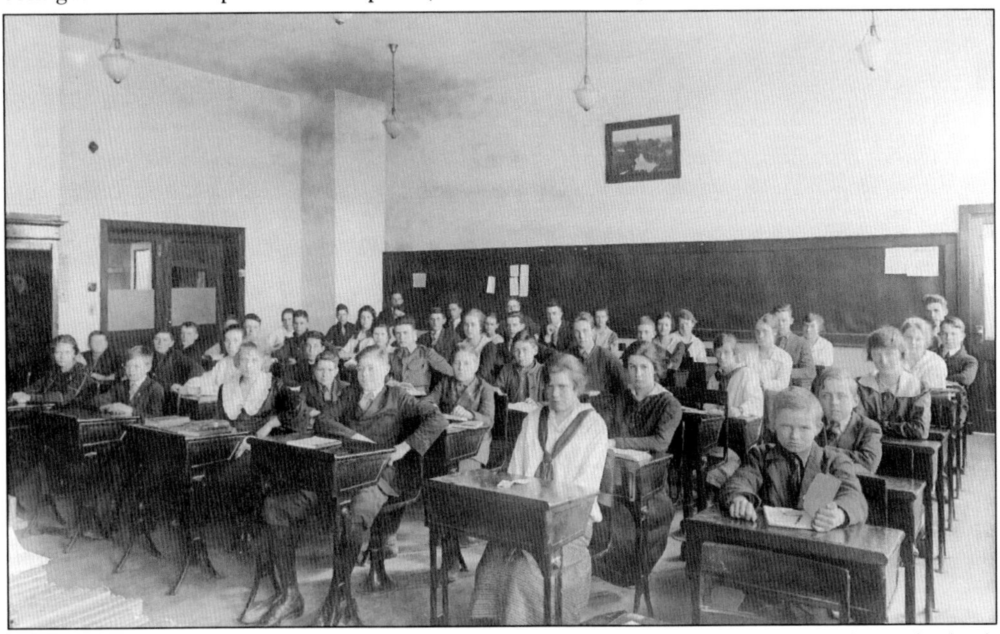

This turn-of-the-century classroom shows a well-lit area with a single photograph on the back wall. The desks have metal bases with wooden tops that open to reveal a small place for storing books, papers, pencils, and other personal items. Formal dress is evident, as are the aisles between rows of desks to give access to the teacher. (NYMHS.)

Two

Evolution to a Company Town

Civil war posed an immediate problem for the textile mills, which were no longer able to obtain cotton required for production. Output slowed. Layoffs increased. Profits fell by more than 35 percent. Although government orders soon arrived, resources remained scarce, and production did not return to normal levels until after the conflict. Despite the tenuous atmosphere, the Walcotts and Campbells placed their resources solidly behind the Union effort. Benjamin Walcott donated a steamboat to the cause. Two of his sons joined the Union army. Samuel Campbell was a leader in the Oneida County War Committee and served as a delegate to the convention that renominated Abraham Lincoln in 1864. On the battlefield, the small village sent off more than 150 men to preserve the nation and end the institution of slavery that its residents had long opposed. Following the conflict, Campbell funded a monument in Glenside Cemetery in 1871. It featured a soldier at ease with the names of the 25 men who made the supreme sacrifice etched into its base.

Following the war, the community continued its evolution into a prototypical "company town." The Walcotts and Campbells insisted on a strict moral code within the village. Alcoholic beverages and profanity were prohibited. Employees and their families were required to observe the Sabbath. Curfew bells rang at 9:00 p.m. when everyone was to be home to rest for the next day's work. Watchmen patroled at night looking for any lights that might be on after 10:00 p.m. when people were supposed to be asleep. Violation of any of these rules was cause for dismissal. Rev. E.J. Kallina explained:

> Mr. Walcott was personally interested in the welfare of all of his men. He was the first manufacturer in this country who reduced the working hours of the day from fifteen to twelve. He was the first who introduced the custom of cash-payment, thus allowing those he employed to purchase their supplies where they could do it cheapest and best. In these and other ways he consulted the best interests of those he had gathered around him and won their sympathy.

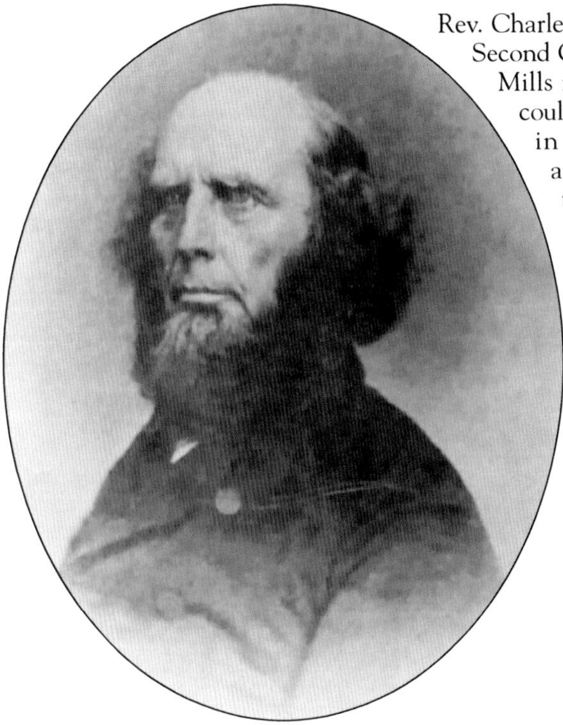

Rev. Charles Grandison Finney, an originator of the Second Great Awakening, preached in New York Mills in the 1830s. His message that salvation could be earned through personal involvement in good causes found ready acceptance among village residents who quickly made the community into a hotbed of radical abolitionist feelings. At least twice during the antebellum period, ministers were dismissed for not being sufficiently antislavery in sentiment. (OCHS.)

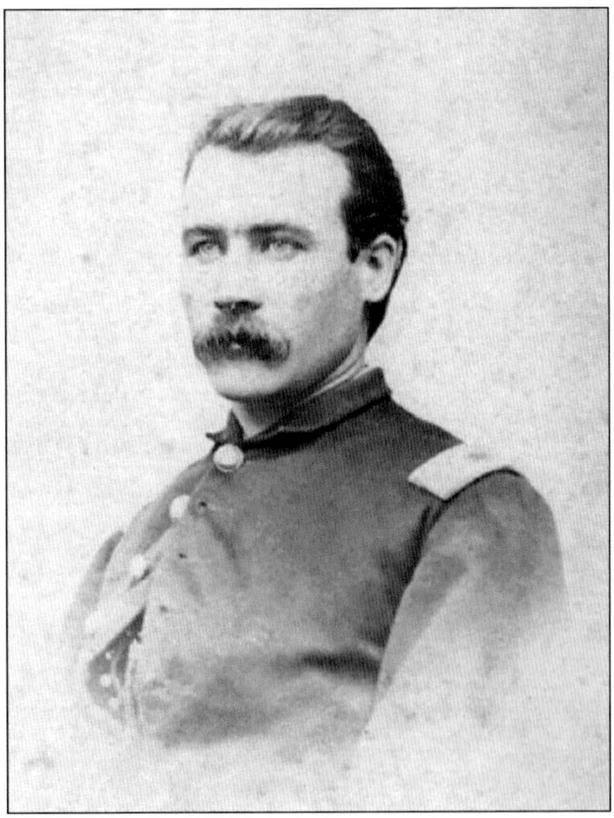

Lt. Andrew Bayne enlisted in the 16th New York Infantry, fought in the First Battle of Bull Run, and won brevet promotions to first lieutenant and captain for gallantry in action. His account of Bull Run, written in a letter home, is an honest description of his own thoughts and actions as well as the scenes he personally witnessed. (Wisnoski.)

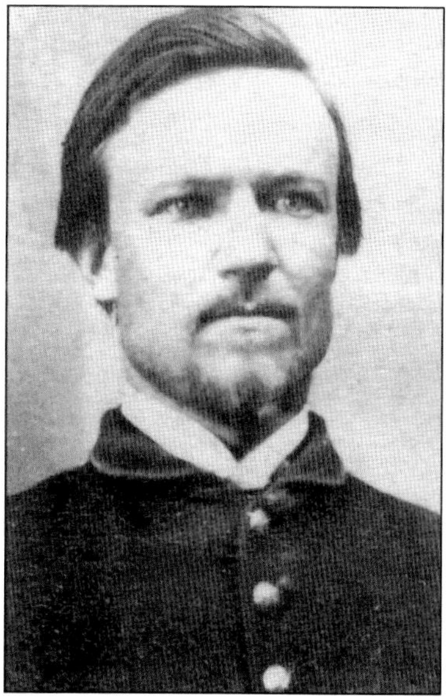

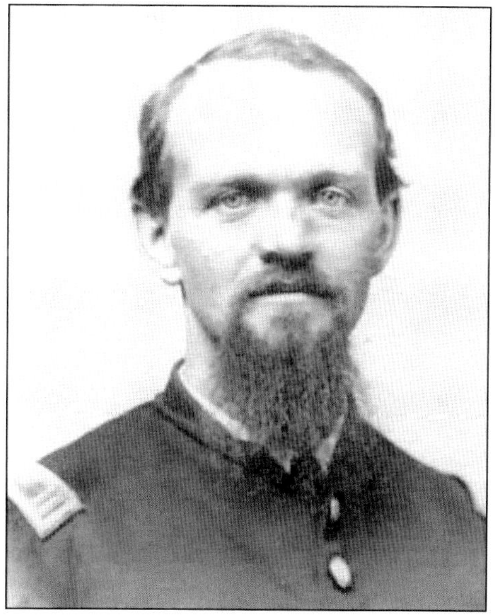

George Fairhead (left), enlisted in the 117th New York Infantry Regiment, was wounded in Virginia in October 1864, and returned home to become an ordained minister. He lived to the age of 99. As the surviving senior officer in the regiment, Capt. David Magill (right) led a successful attack on Fort Fisher in January 1865. Although losing a leg in the battle, he was brevetted a major for his heroic conduct. (Wisnoski.)

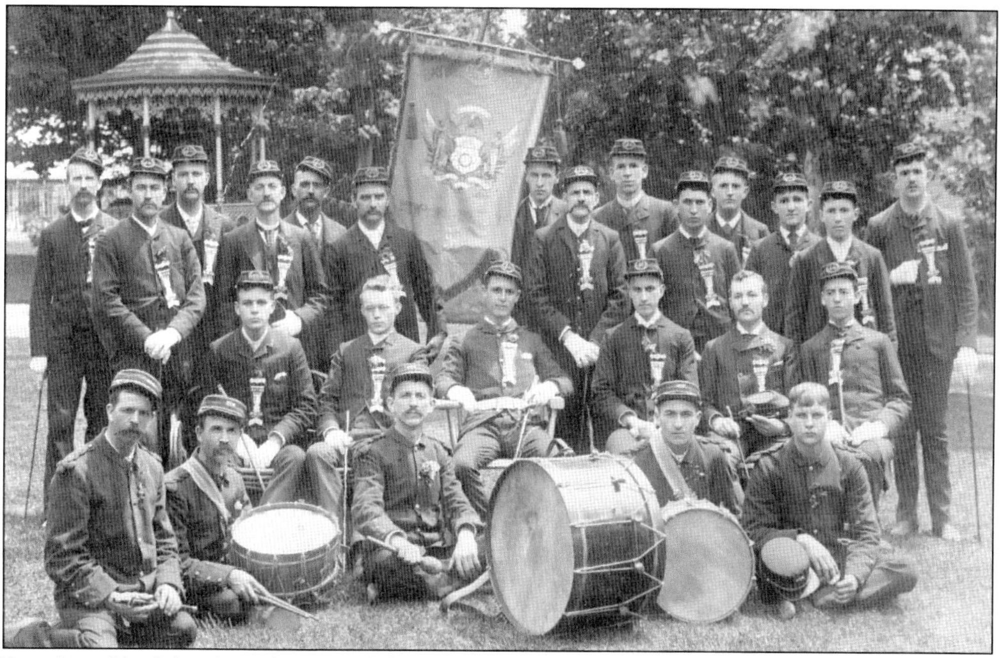

Ross Post 31 of the Grand Army of the Republic formed in the village by veterans of the Civil War. Pictured here on Memorial Day in 1887 is Bayne Post of the Sons of Union Veterans and the five-man New York Mills Fife & Drum Corps. Both organizations carried out civic activities that included everything from commemorating national holidays to organizing lectures and charitable events. (OCHS.)

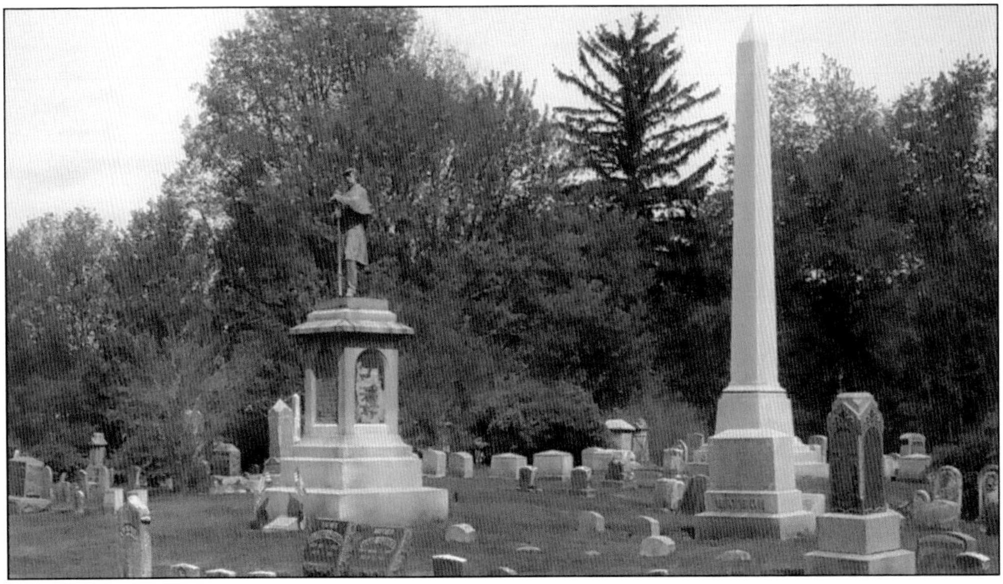

Donated by the Campbell family, the statue of an infantryman "at ease" in Glenside Cemetery was the first Civil War monument erected in Oneida County. Listed on the sides of the memorial are the names of 25 residents who gave their lives during the war. It is located in front of the Campbell family obelisk at the entrance to the cemetery. (Authors' collection.)

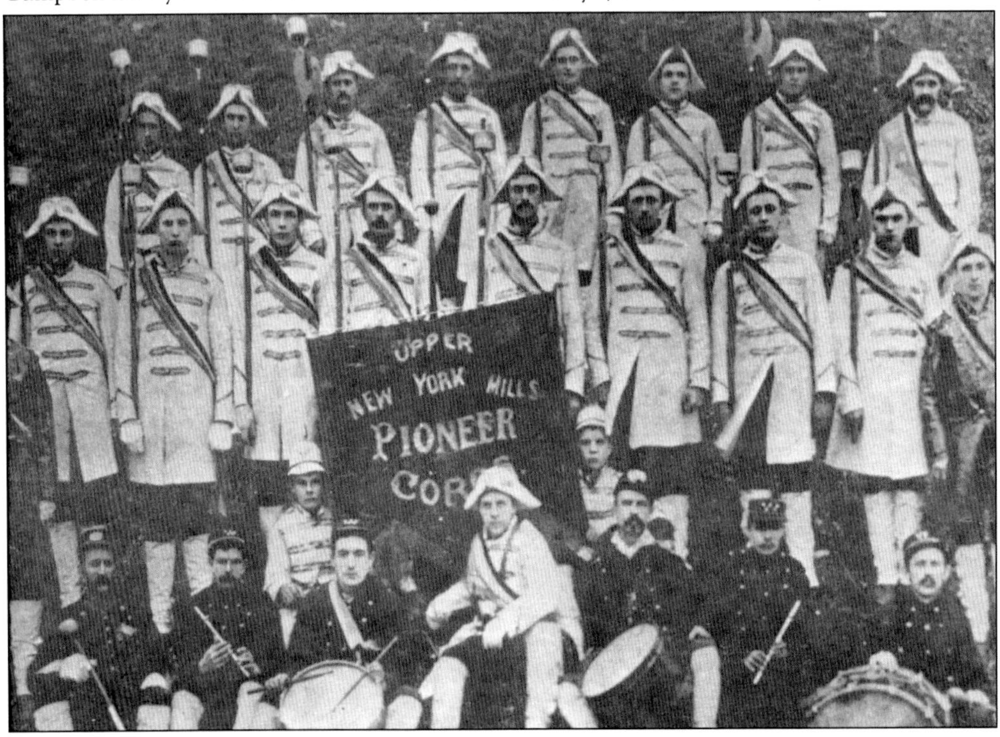

Pictured here in 1884, the Upper New York Mills Pioneer Corps dressed in Colonial-style uniforms to march in patriotic and community parades. At the bottom of the photograph are members of the New York Mills Fife & Drum Corps, which similarly played for community and patriotic occasions. (NYMHS.)

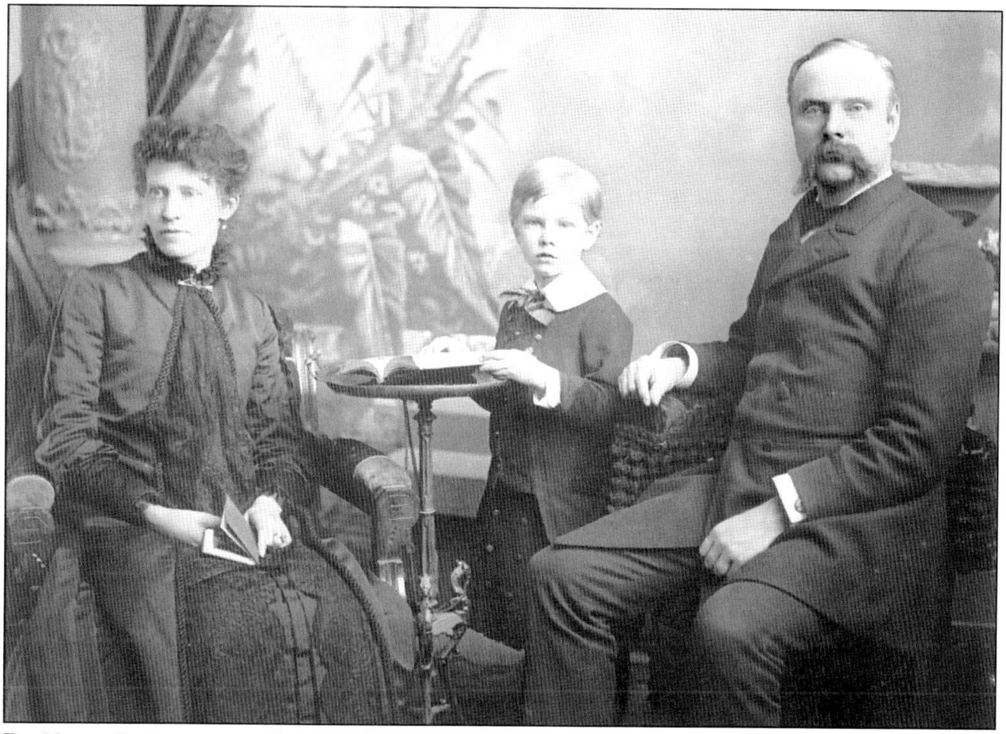

Dr. Henry R. Porter poses here with his wife, Lottie Viets Porter, and their son Henry V. Porter. Dr. Porter, a graduate of Georgetown University, was the sole surviving surgeon with the 7th Cavalry when Gen. George A. Custer was defeated at the Little Big Horn battle in 1876. He gained praise for his professional conduct with the wounded while under fire. Porter Street is named for his family. (SHSND.)

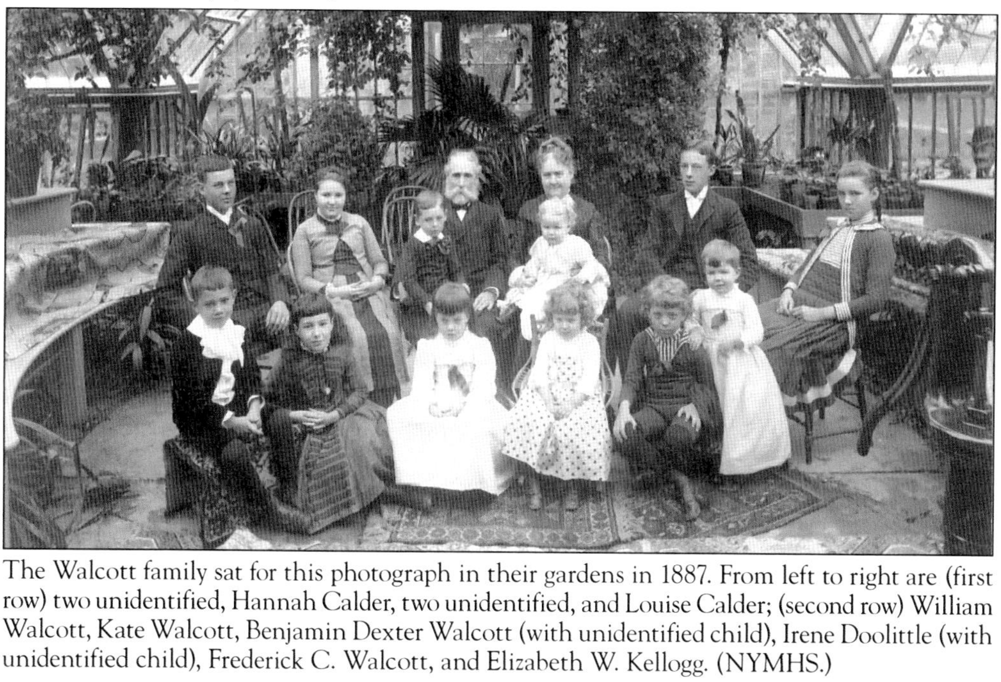

The Walcott family sat for this photograph in their gardens in 1887. From left to right are (first row) two unidentified, Hannah Calder, two unidentified, and Louise Calder; (second row) William Walcott, Kate Walcott, Benjamin Dexter Walcott (with unidentified child), Irene Doolittle (with unidentified child), Frederick C. Walcott, and Elizabeth W. Kellogg. (NYMHS.)

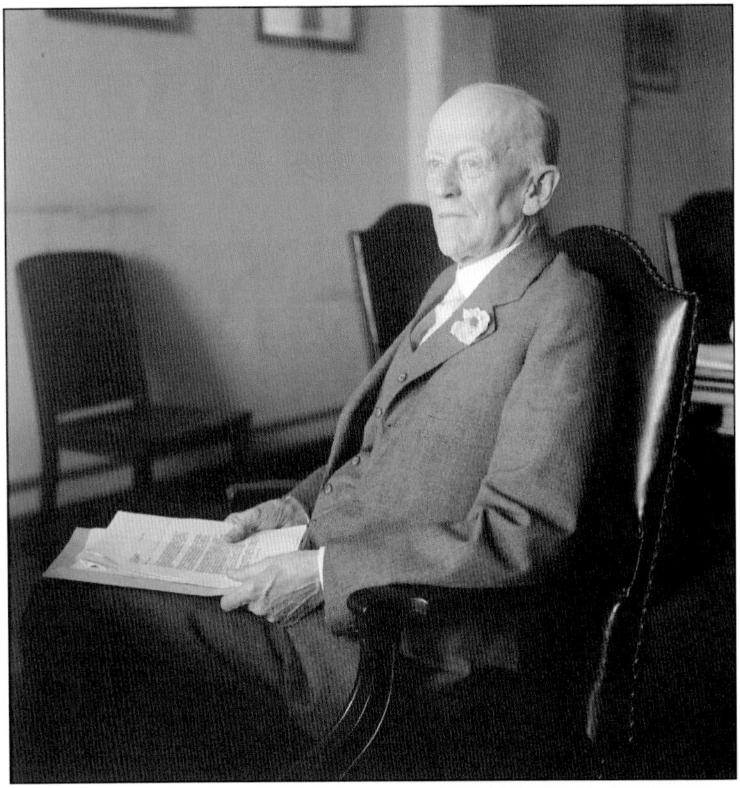

Charles Doolittle Walcott was born in New York Mills in 1850. He became director of the US Geological Survey in 1894. From 1907 until his death in 1927, he served as secretary of the Smithsonian Institution. He was instrumental in founding the Carnegie Institution in Washington, DC, the Freer Gallery of Art, the National Research Council, and the National Advisory Council for Aeronautics. (Library of Congress.)

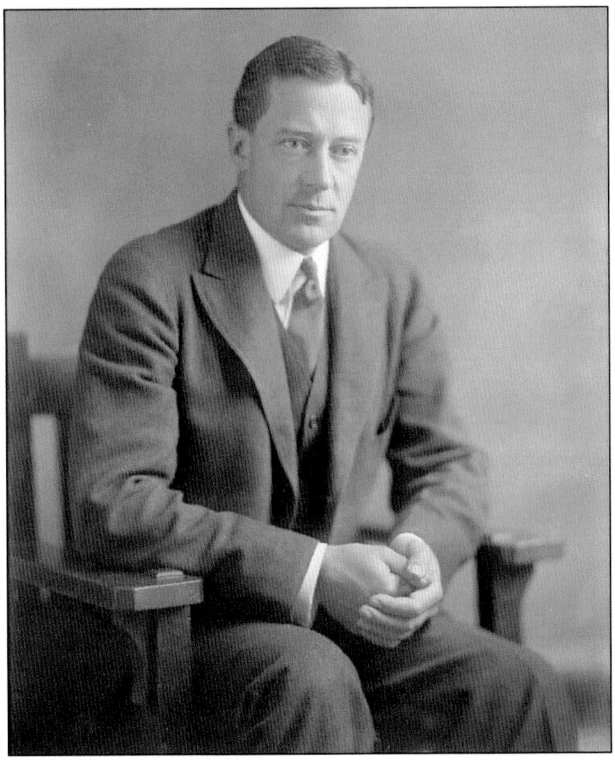

Frederick Colin Walcott directed the relief missions in Belgium and Poland from 1915 to 1916. He was an advisor to President Wilson on Polish and Russian affairs at Versailles. Later, after moving to Connecticut, he was elected to the US Senate, served on the board of regents of the Smithsonian Institution, and participated in the Manhattan Project and other scientific programs during World War II. (Library of Congress.)

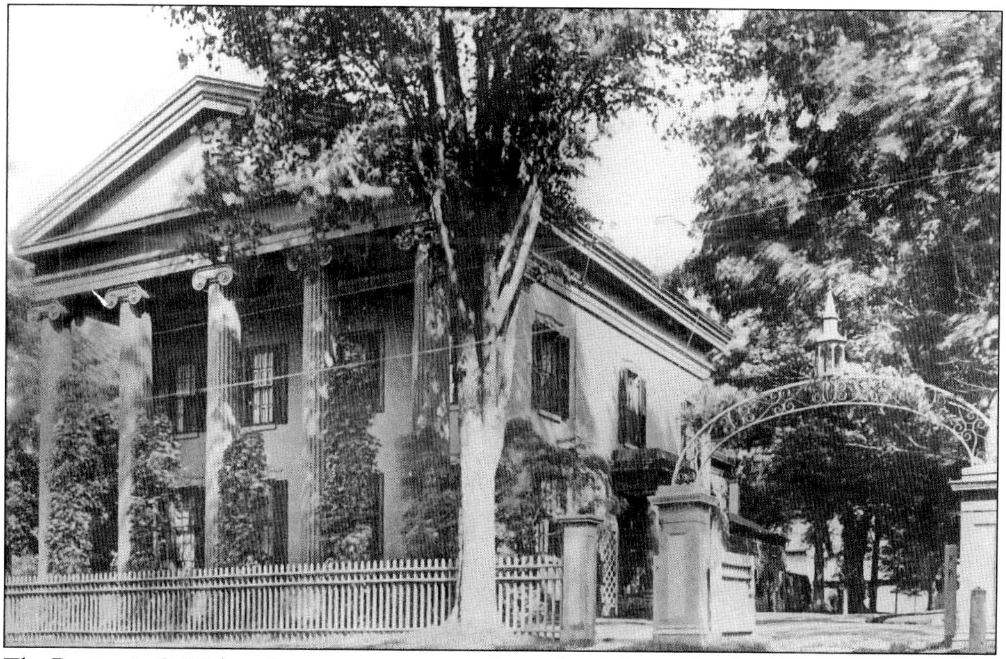

The Benjamin S. Walcott home was constructed in a Greek Revival style. The three-story temple building included an attached portico and five Ionic columns fronting the structure. Walcott was the son of the original mill founder. It was he who was largely responsible for building the groundwork of the successful textile industry in the village. (OCHS.)

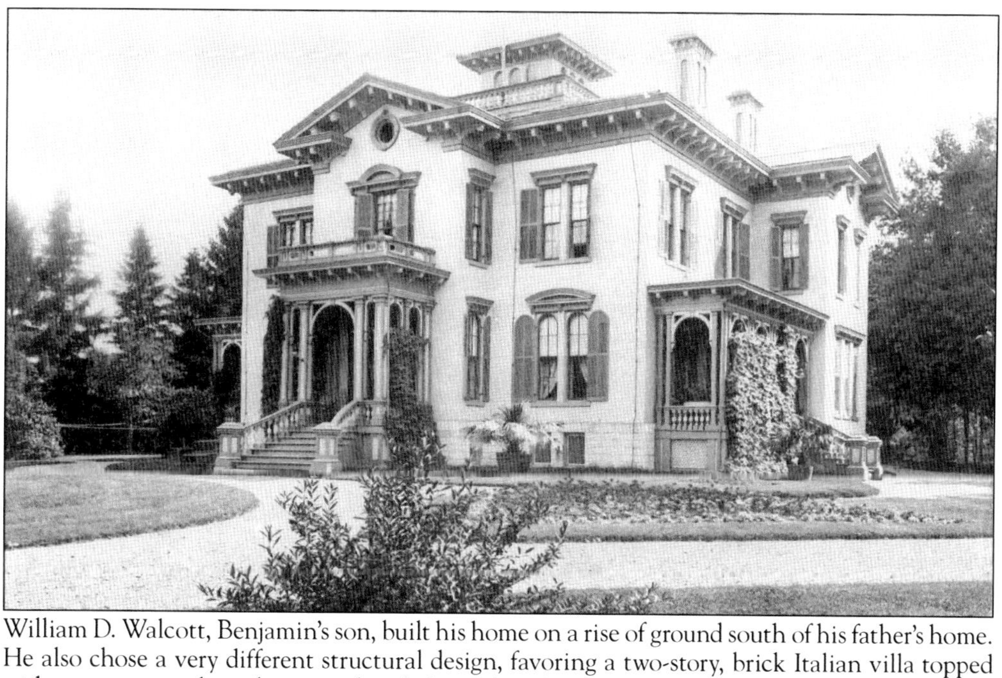

William D. Walcott, Benjamin's son, built his home on a rise of ground south of his father's home. He also chose a very different structural design, favoring a two-story, brick Italian villa topped with a square cupola and surrounding balustrade. The ornate result contrasted sharply with his father's simpler design. It is believed the home was painted a faun color with chocolate brown eaves, porches, windows, and shutters. (NYMHS.)

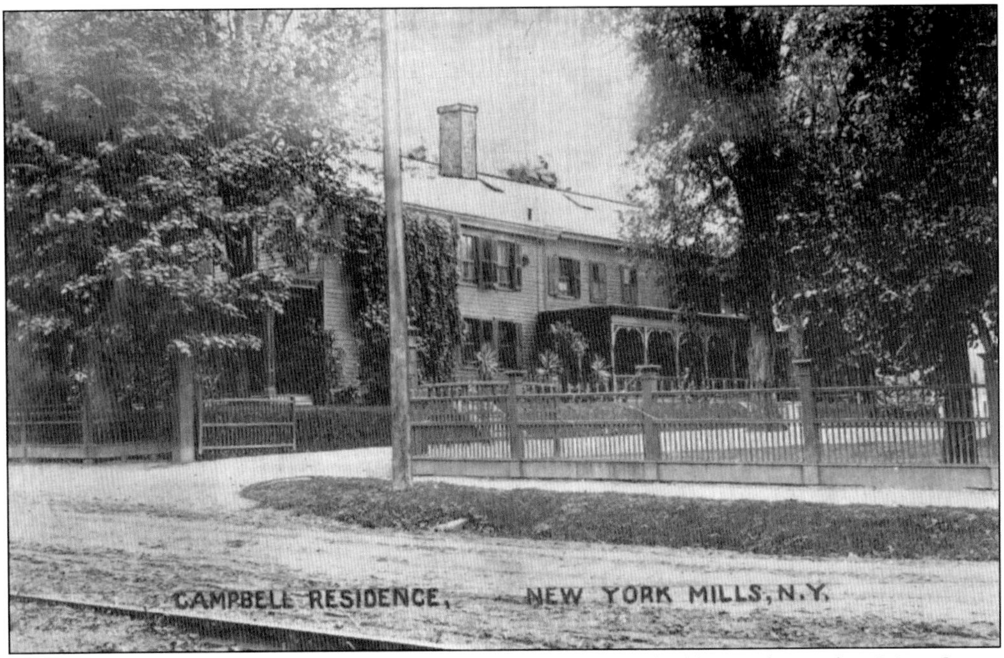

The Samuel Campbell home was a sprawling complex of several additions located on Main Street in the Upper Mills. Campbell was well known for his philanthropy and community activities, including serving a term as supervisor of the Town of Whitestown and attending the 1864 Union Party National Convention in Baltimore that renominated Abraham Lincoln. (NYMHS.)

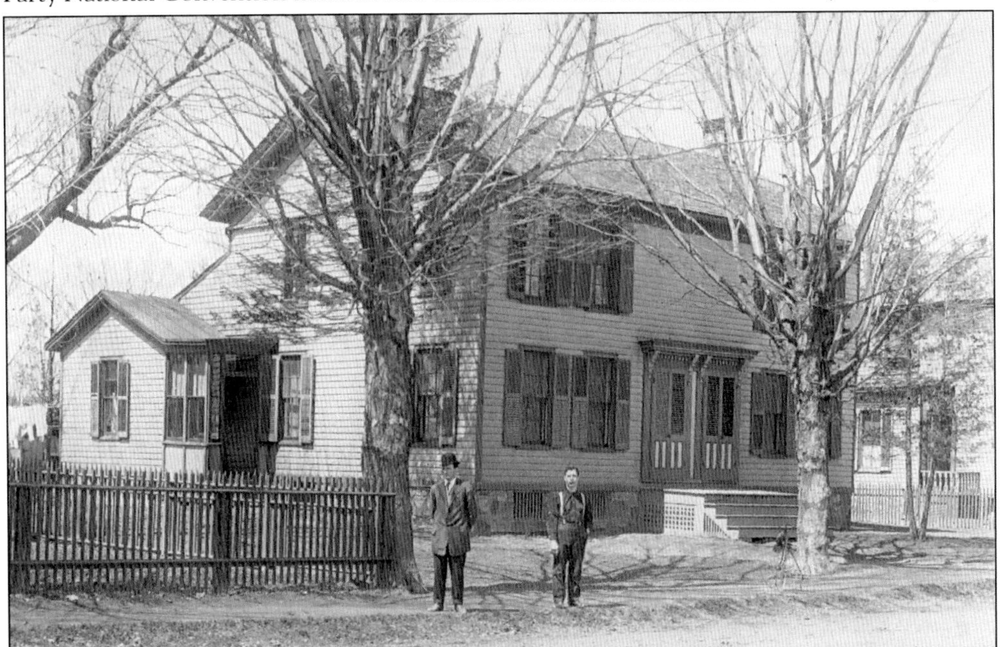

The Charles Roberts residence was a good example of the kind of home a relatively prosperous middle class manager could expect to own. Most of the superintendents and other members of upper management owned their own homes, although the company maintained some better housing that was available to foremen and other management officials. (OCHS.)

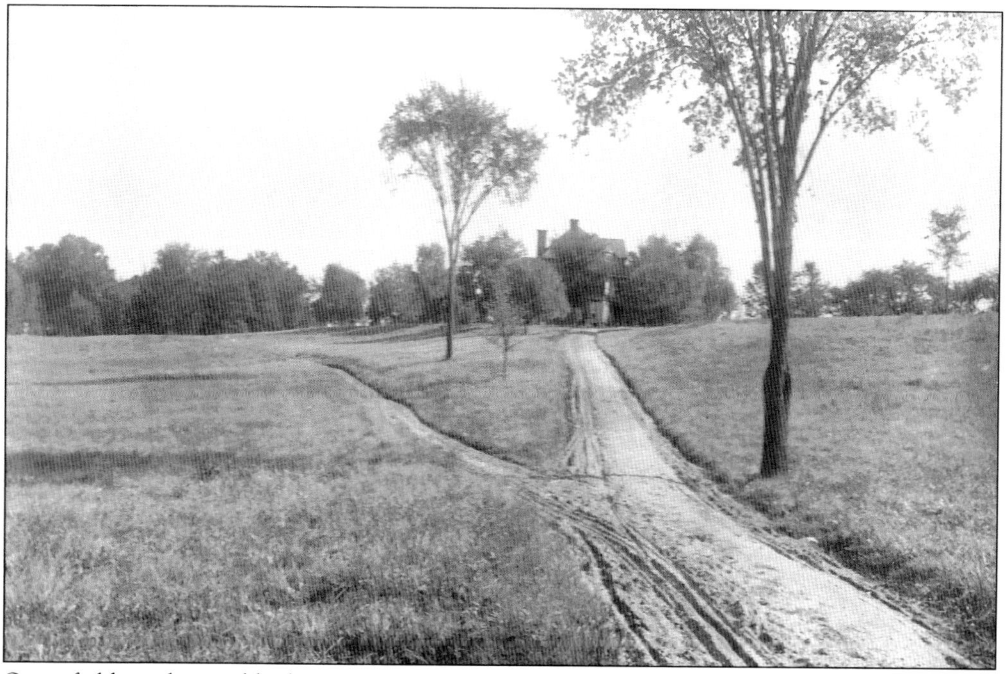

Open fields and a muddy dirt road led up this small hill to a clump of trees surrounding the residence of Samuel Campbell's son-in-law George H. Warner, superintendent of the Lower Mills. This location later became St. Mary's Street leading up the rise to St. Mary Our Lady of Czestochowa Catholic Church. (OCHS.)

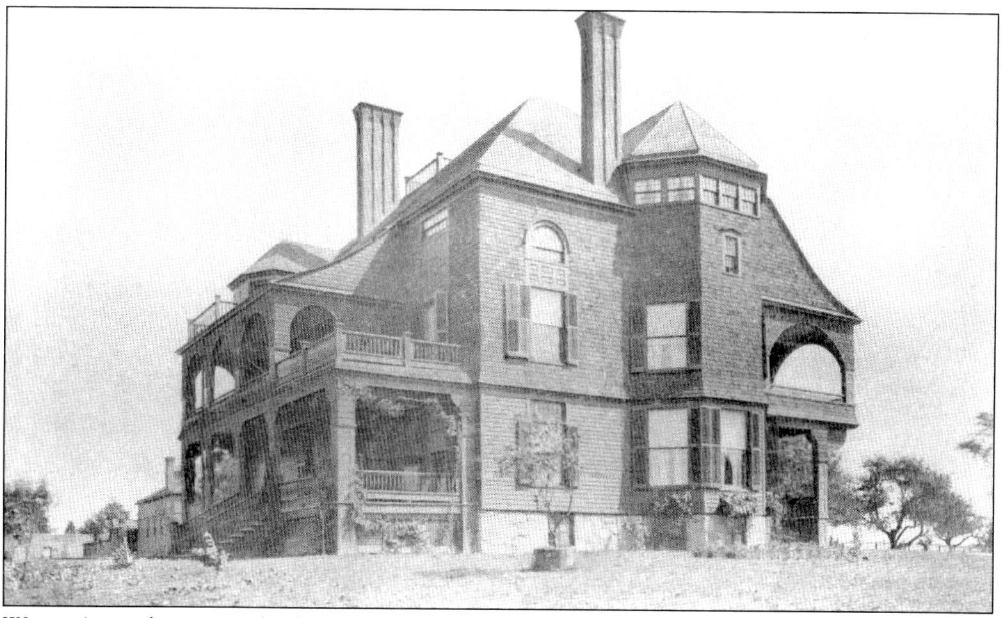

Warner's residence was built in Shingle style with an unusual asymmetric shape and Tudor chimneys. It contained a basement, two floors, and an attic. The first floor had three large rooms, a dining room and a kitchen, while the second floor contained bedrooms, each room having its own fireplace faced with yellow tile and an oak mantelpiece. This later became the rectory for St. Mary's Parish. (UPL.)

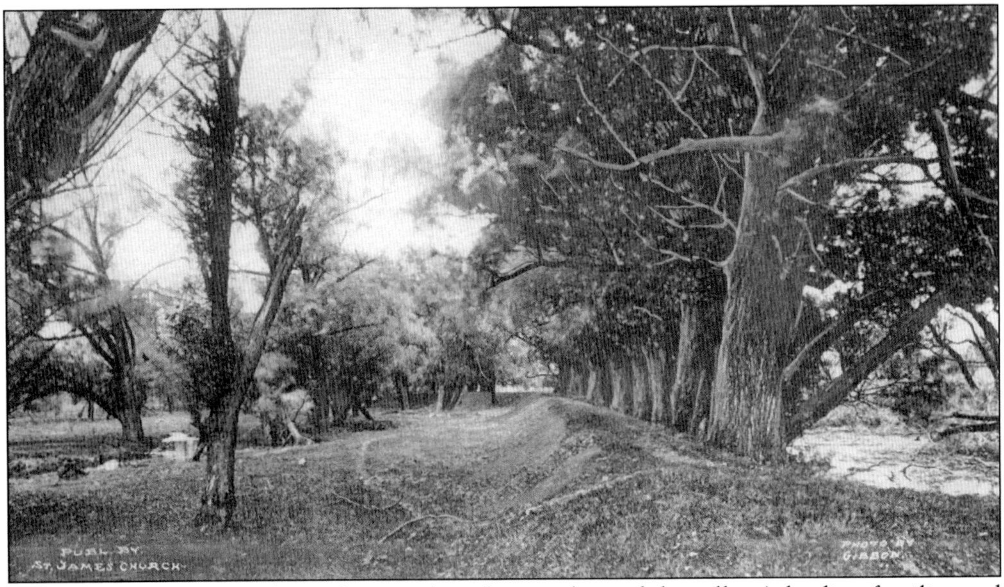

The Willows was a popular gathering place for members of the village's leading families and management officials from the mills. It was located on the Campbell estate in the Lower Mills. With mature trees, ponds, and open fields, it was ideal for picnics and group outings when not in use as a grazing area of the Campbell livestock. (Pytko.)

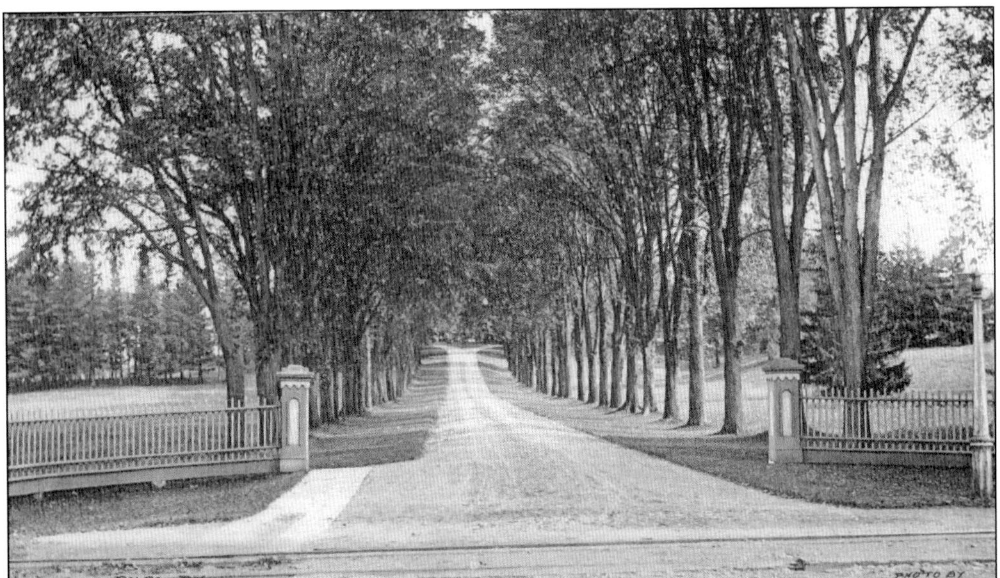

When the older trees on the Campbell estate were felled by a major storm, the area was cleared, new trees were planted, a more permanent road was installed, and the former Willows area became the new front entrance to the Campbell estate. It retained this general look in later years, eventually becoming the entrance to the Twin Ponds golf course. (NYMHS.)

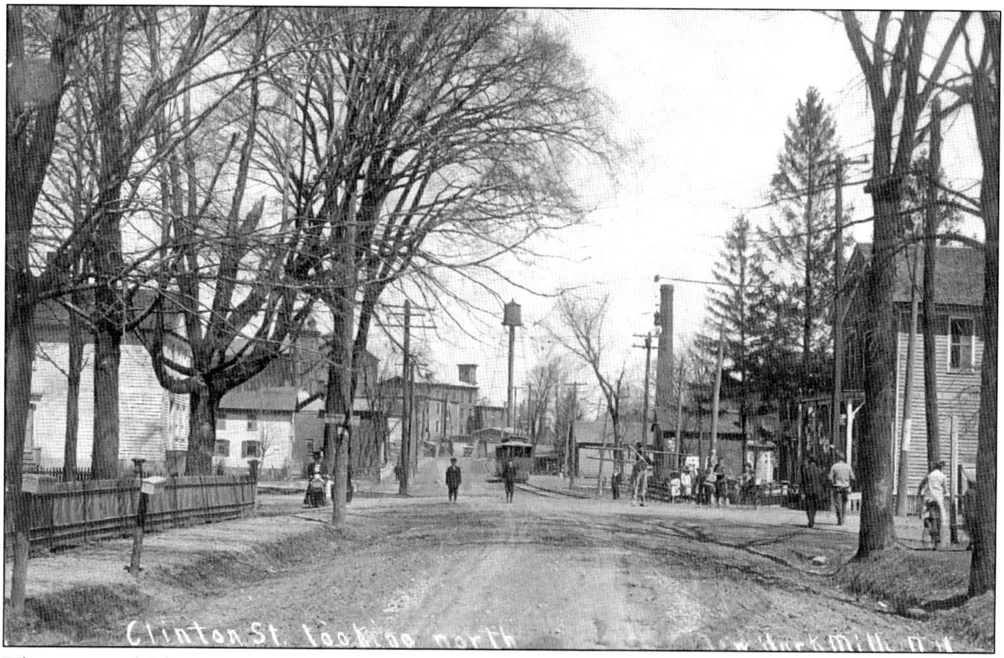

This view of Clinton Street, looking toward the northeast, shows McCarthy's Corner where Clinton Street intersects with Henderson and New Hartford Streets. Named for the McCarthy store seen at the right, raised but unpaved walkways with mailboxes can be see along the roadsides with the Upper Mills in the distance. (NYMHS.)

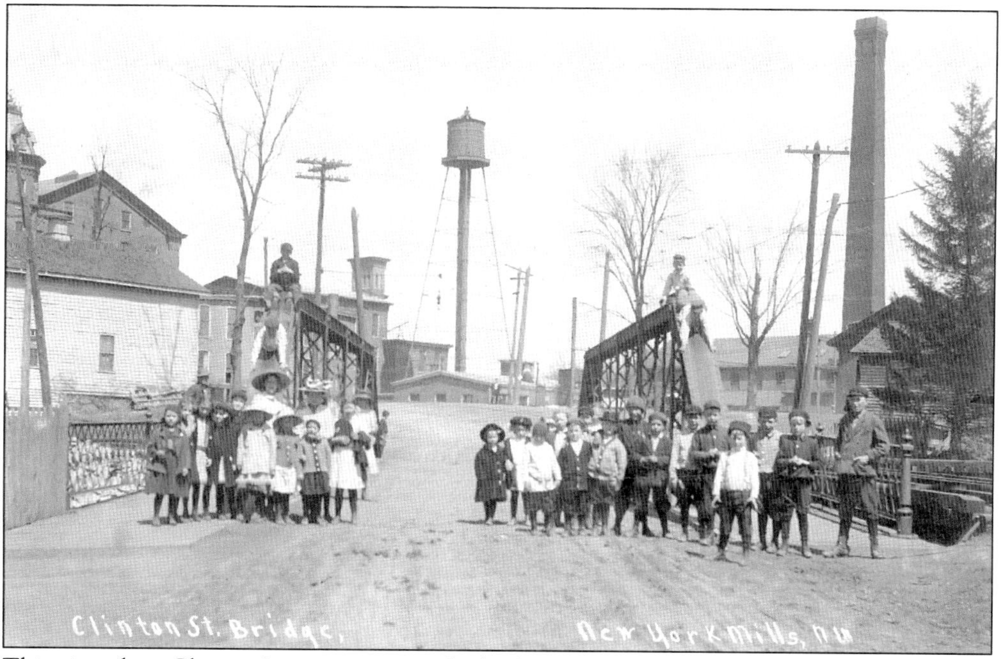

This view along Clinton Street centers on the bridge over Sauquoit Creek. The dye house facility is to the right where the smokestack is located, while to the upper center with the water tower is the location of mills No. 3 and 4. Note the children's dress, which appears much like that in the earlier school portrait. (NYMHS.)

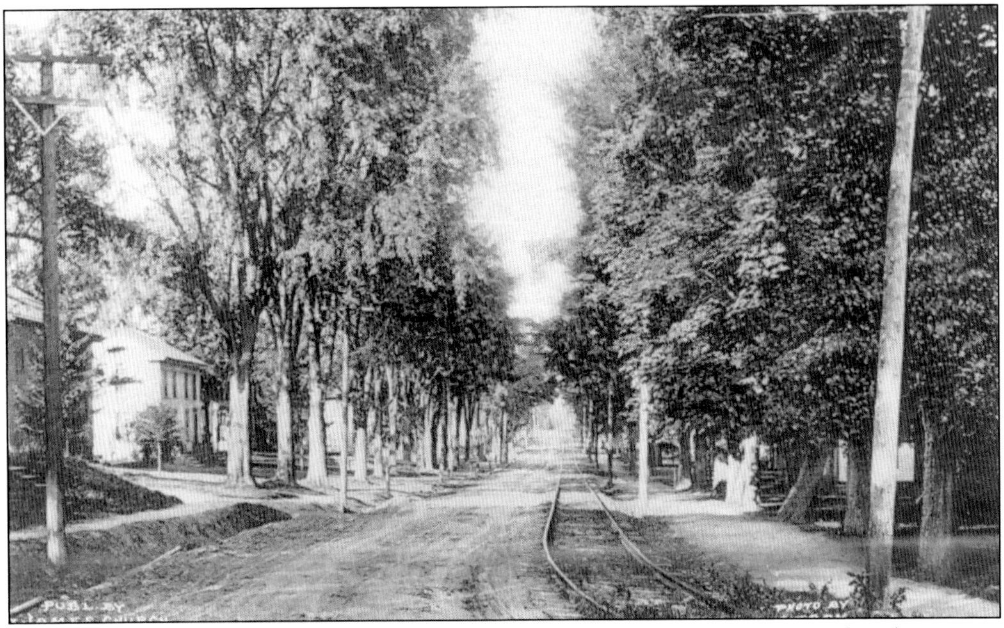

Looking north along Main Street from where the Landmark Apartments are today, this view is of a dirt street with raised dirt walkways on both sides and trolley tracks to the right. The homes are all located back from the street, separated by plentiful shade trees lining both sides of the roadway. Note also that electric lines have made an appearance. (NYMHS.)

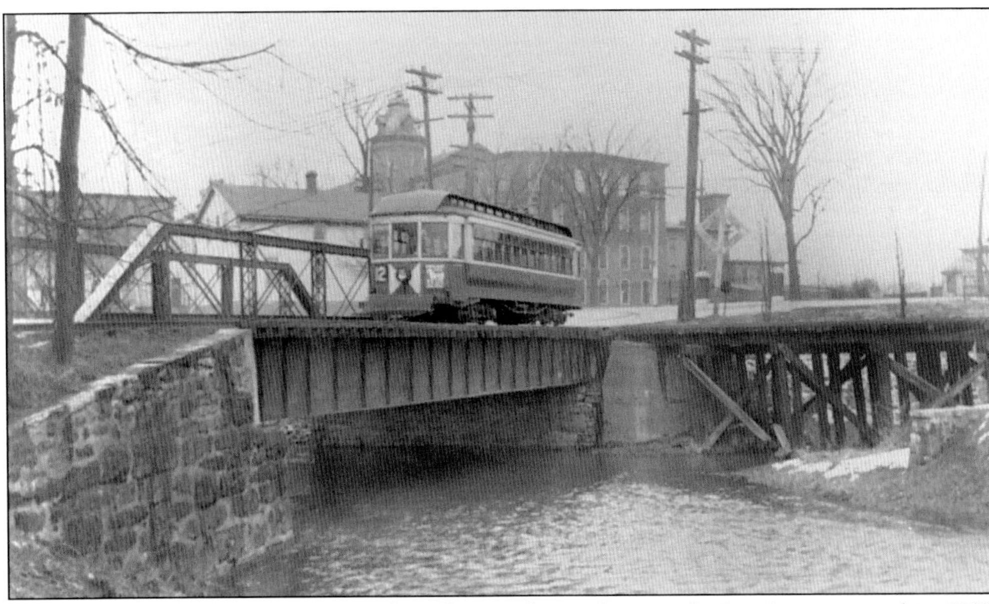

The first streetcars serving New York Mills were horse-drawn vehicles that appeared in 1887. This electric version is seen with the No. 4 mill (today No. 3) in the background. To the right is a railroad trestle used by the freight trains that served the mills. The tracks ran through the village, passing near each mill and connecting to the main east-west line of the New York Central. (NYMHS.)

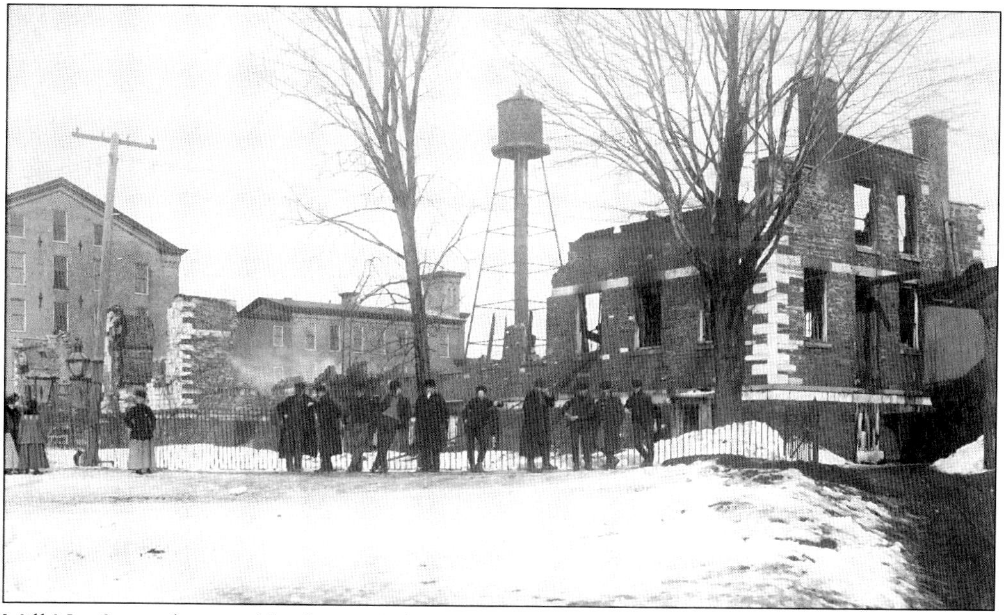

Mill No. 3 was destroyed by fire on February 29, 1904, leaving little more than portions of the original shell. After the debris was removed, the space was used to construct a smaller appendage to the remaining No. 4 mill and, later still, a parking lot. Over time, without a No. 3 mill, residents began to refer to the No. 4 as "No. 3." (NYMHS.)

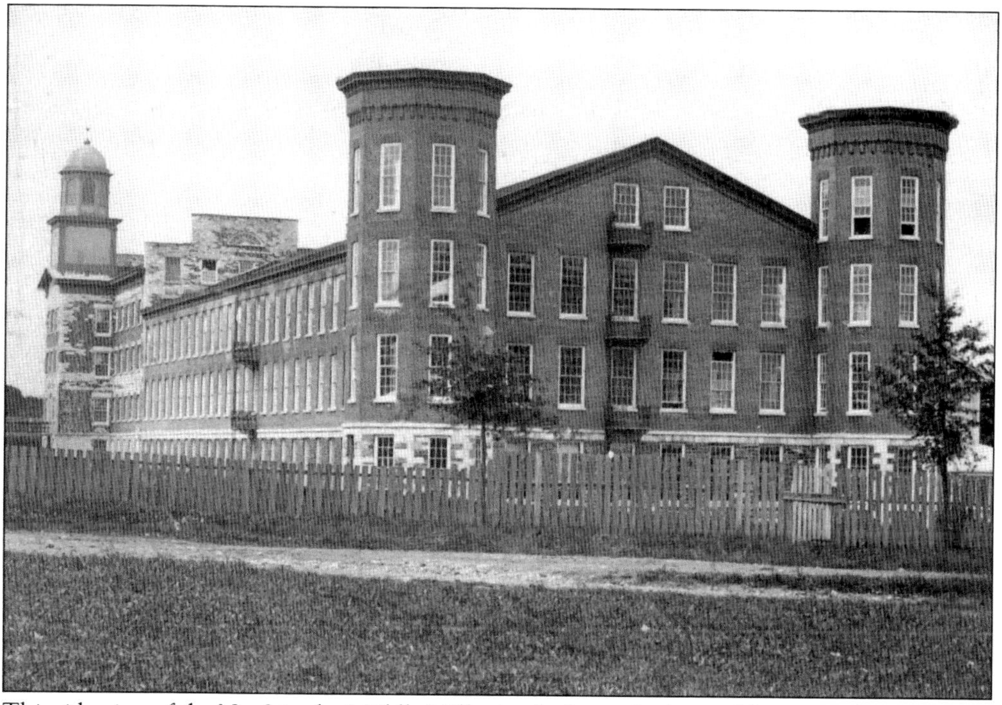

This side view of the No. 2 in the Middle Mills clearly shows the later additions to that structure, the red brick and functional design in the foreground contrasting sharply with the original stone Federalist facade on the left. The nearest portion with the octagonal towers was completed in 1884. (OCHS.)

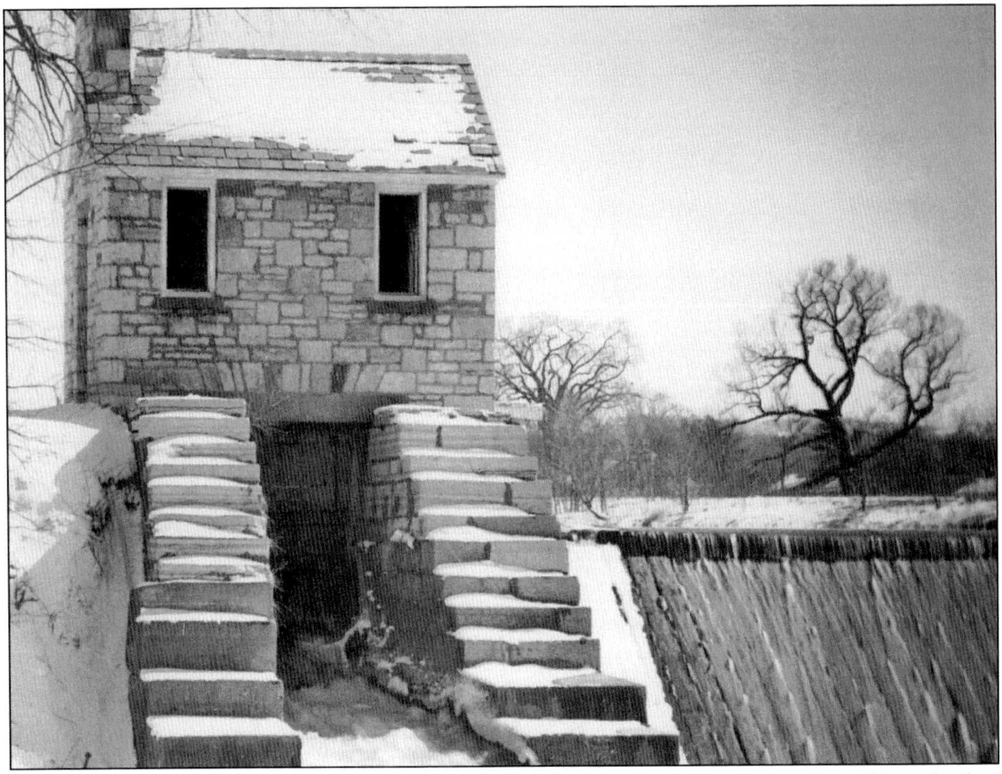

This gatehouse behind the No. 2 was constructed of stone at the same time as the mill to which it provided power. The building was used by the gatekeeper to raise or lower the door allowing water into the sluice. Sauquoit Creek took its name from a Brotherton and Stockbridge Indian word meaning "short and rapid," a good indication of why it was chosen as a mill location. (NYMHS.)

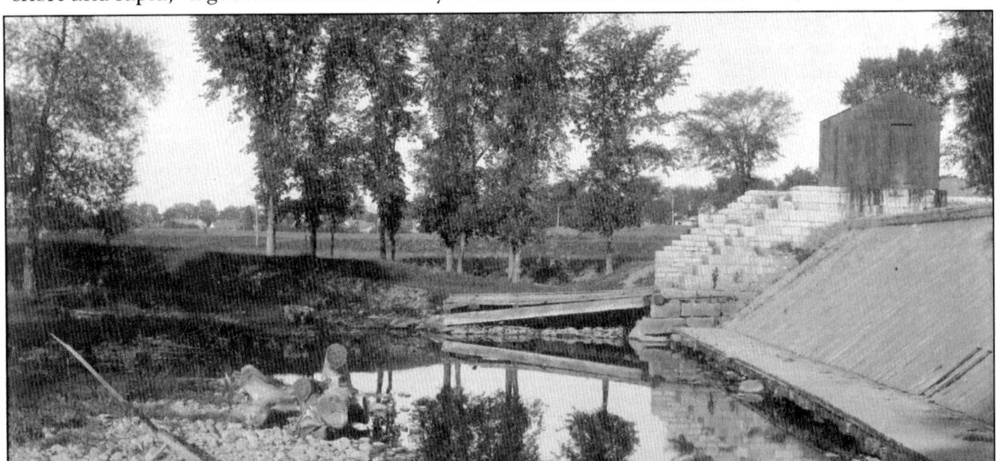

This 1899 photograph shows one of the dams used to regulate the flow of water to the mills. Before modern environmental regulations, waste from the dye house was dumped into the creek. Children who swam or waded in the waters during the hot summer months had to be careful or they might go home with blue, green, red, or other coloring depending on the color dumped that day. (OCHS.)

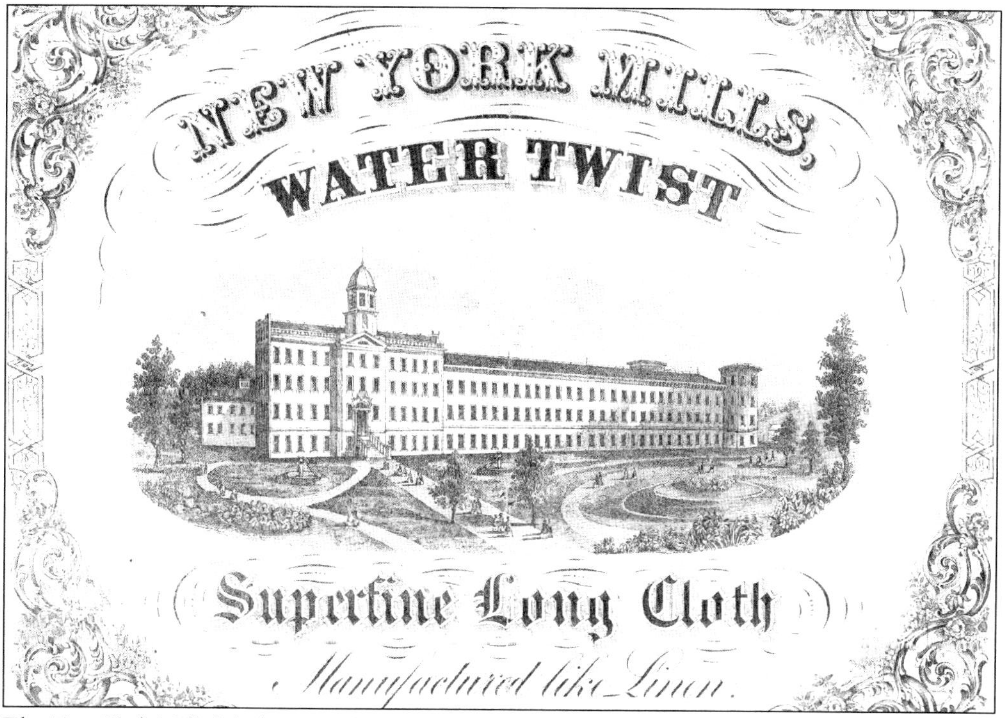

The New York Mills label was a valued commodity throughout the 19th century. As early as 1828, testimony given to the US House of Representatives Committee on Manufacturers stated, "No other factory in the United States [was] weaving cottons as fine as those of the New York Mills. Their cloth was said to be the standard for fine fabrics wherever produced." (Pytko.)

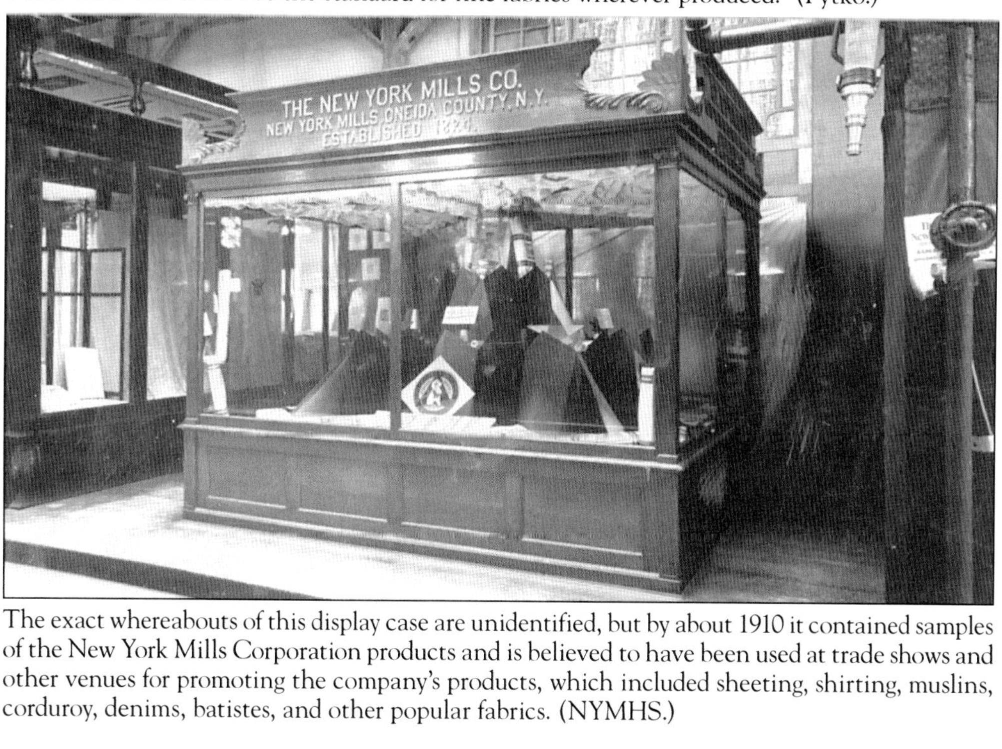

The exact whereabouts of this display case are unidentified, but by about 1910 it contained samples of the New York Mills Corporation products and is believed to have been used at trade shows and other venues for promoting the company's products, which included sheeting, shirting, muslins, corduroy, denims, batistes, and other popular fabrics. (NYMHS.)

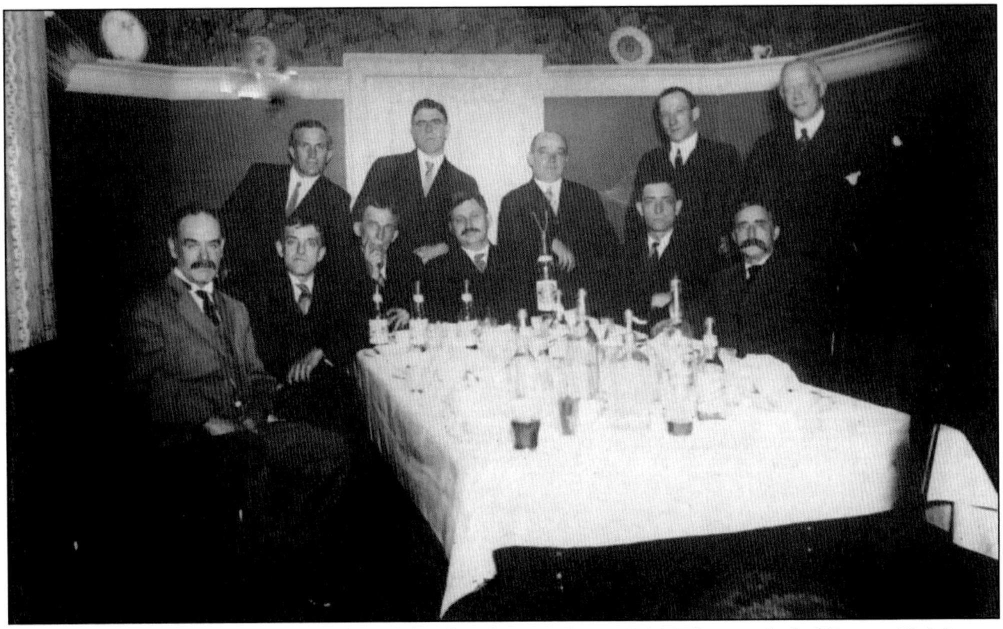

Officials from the textile mills gathered for this dinner about 1914. They included Ernest H. Moore, chief power engineer (standing at left); "Chip" Tabor, boss of the steamfitters and plumbers (seated third from left); John W. Simpson, mechanical superintendent (seated fourth from left); Dennis Shannon, a superintendent (seated fifth from left); and ? McPherson, boss of the machinists (seated sixth from left). (NYMHS.)

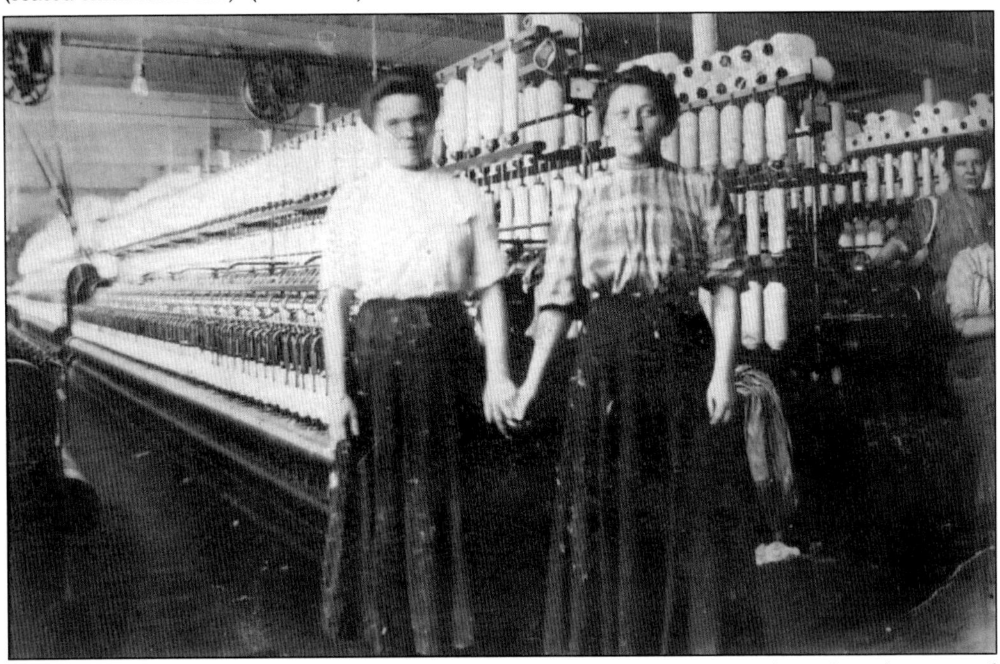

These unidentified female employees pose in the weaving room. Note the calico dresses and the short or rolled-up sleeves, which were important for safety so they did not get caught in the machinery. Cotton fed into the machines was turned into thread that appears on the spools above. Each weaver would be responsible for one bank of these machines. (NYMHS.)

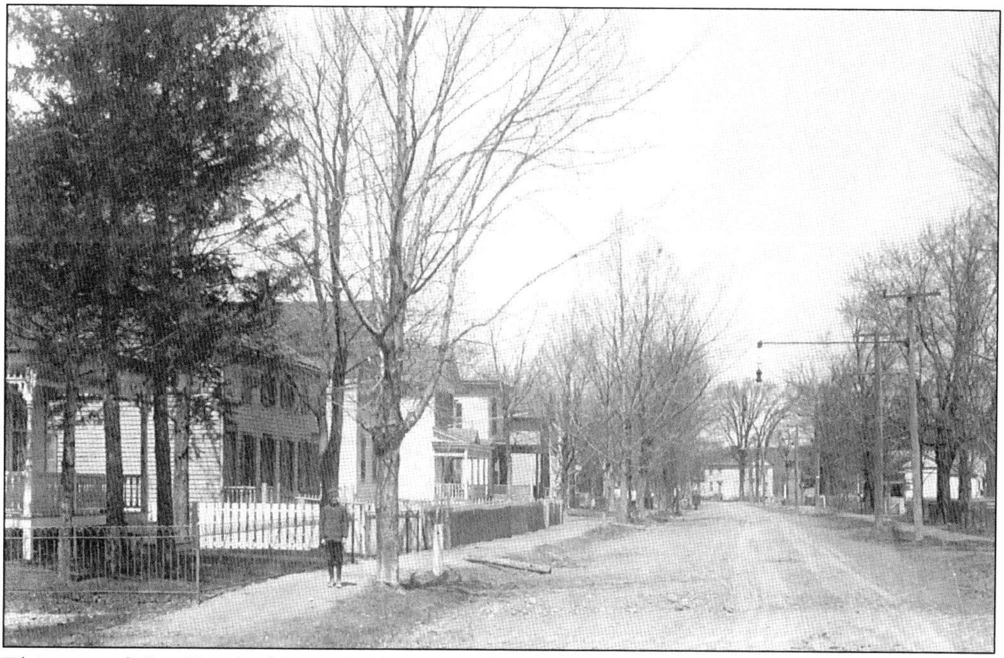

This view of New Hartford Street looking roughly north toward McCarthy's Corner illustrates the typical wood frame construction of middle-class and upper-middle-class houses in the village in the beginning years of the 20th century. Also depicted clearly are the wooden slat fences found about many of the yards. (OCHS.)

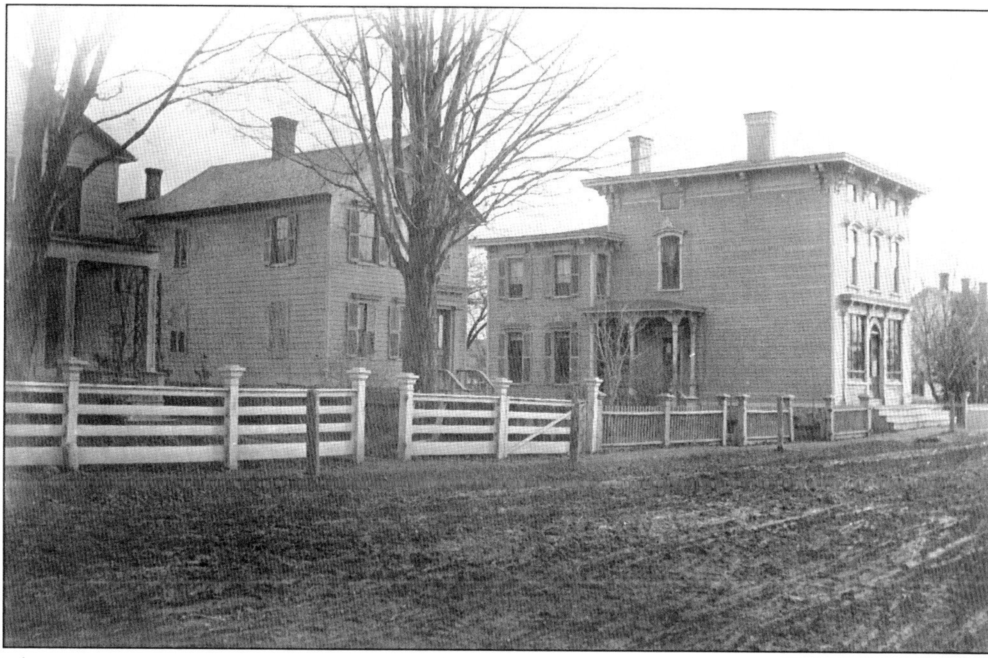

The Fraser Store (on the right), located at 507 Main Street in the Upper Mills, was typical of the growing number of independent merchants springing up by the beginning of the century. Larger and more ornate than most, it contained the store on the ground level as well as living space in the rear and on the second floor, some of which could also be rented out. (OCHS.)

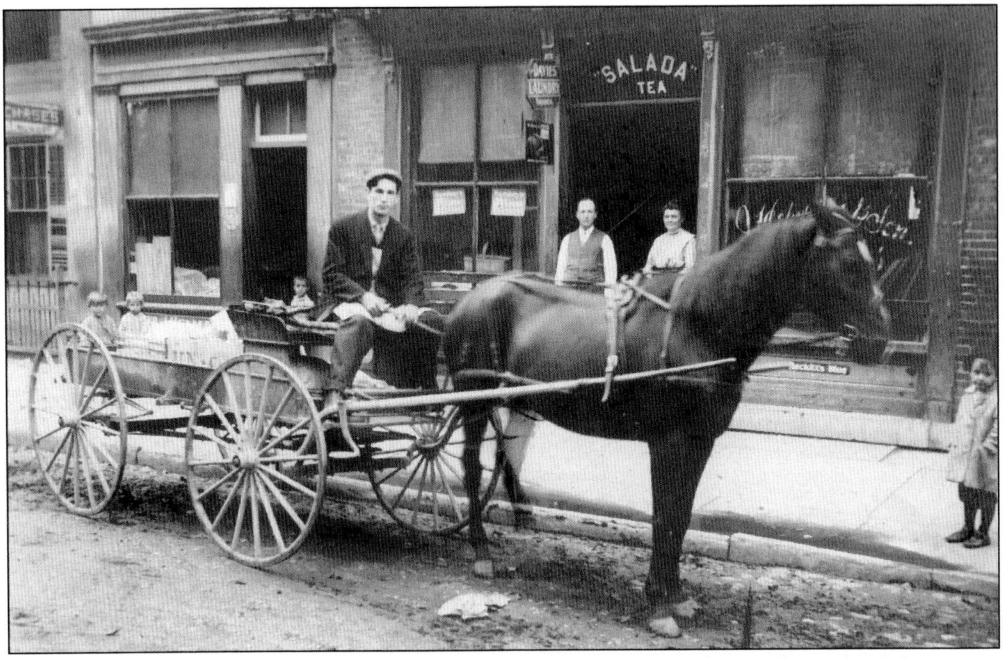

Many of the middle-class entrepreneurs who emerged in the village during the early 20th century began by selling vegetables, dairy products, hardware, or other commodities from wagons like this one seen in front of the "Brick Block." More established merchants with their own stores also used wagons to deliver lunches to the mill workers. (NYMHS.)

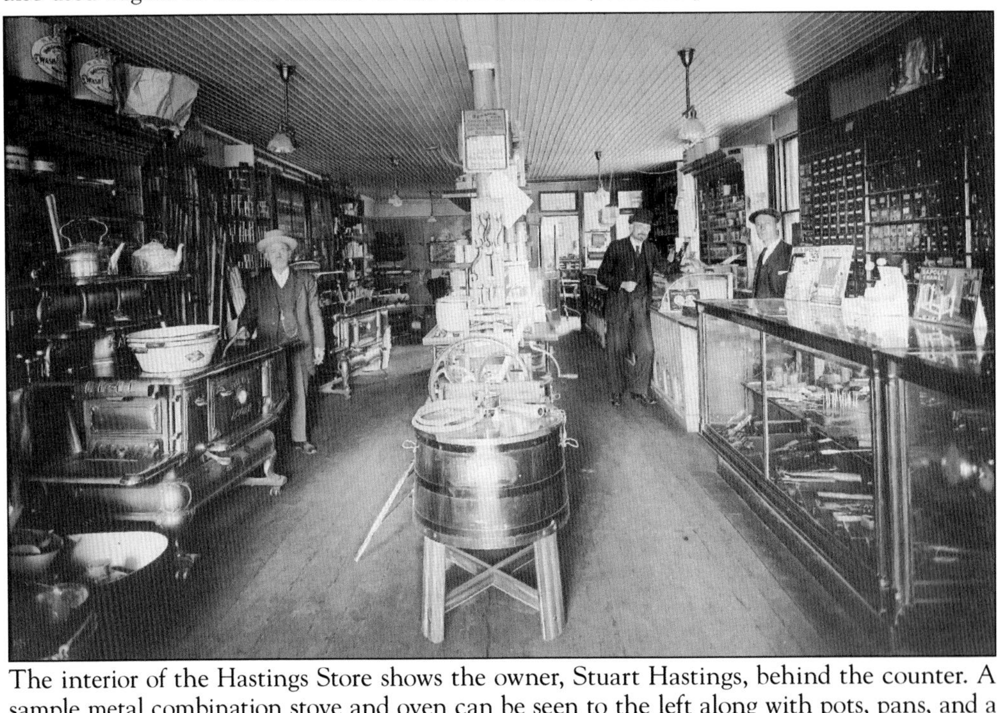

The interior of the Hastings Store shows the owner, Stuart Hastings, behind the counter. A sample metal combination stove and oven can be seen to the left along with pots, pans, and a variety of other merchandise. To the right is a large glass display case. The walls are lined with shelves and drawers for nails, screws, and other small items. (NYMHS.)

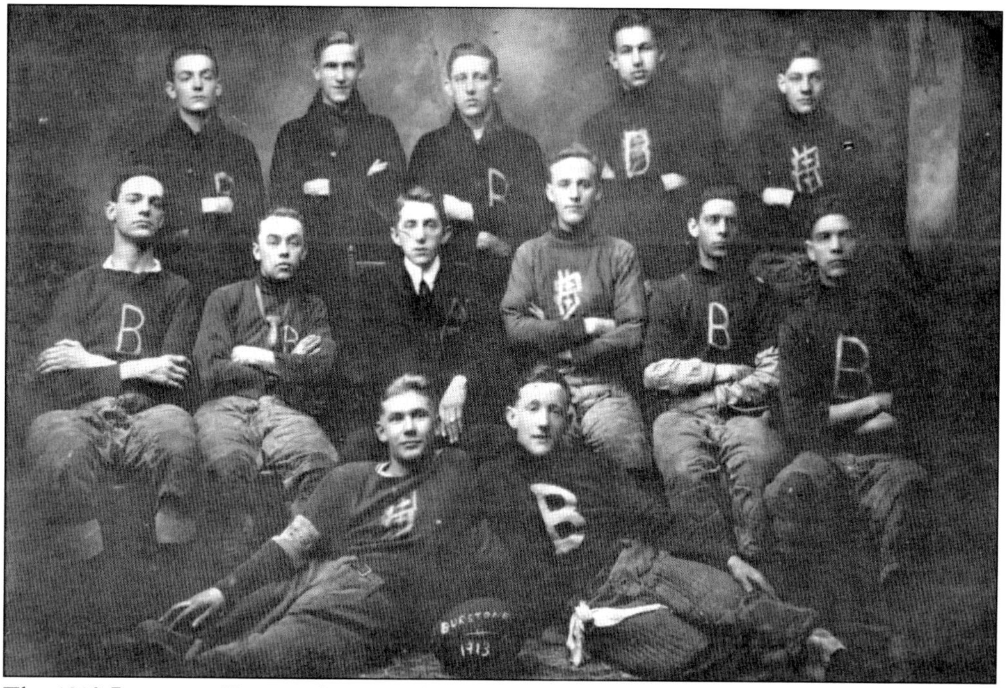

The 1913 Burrstone Terriers championship football team from the Upper Mills featured, from left to right (first row) George Roberts and Laurence Watkins; (second row) Happy Stressel, Joe ?, Alex Alexanders, Joe Simpson, Pete Stockton, and Tom Stockton; (third row) Earl Stressel, Joe Robellard, Snits Combelleck, Charlie Corsette, and Morris Atkenson. (OCHS.)

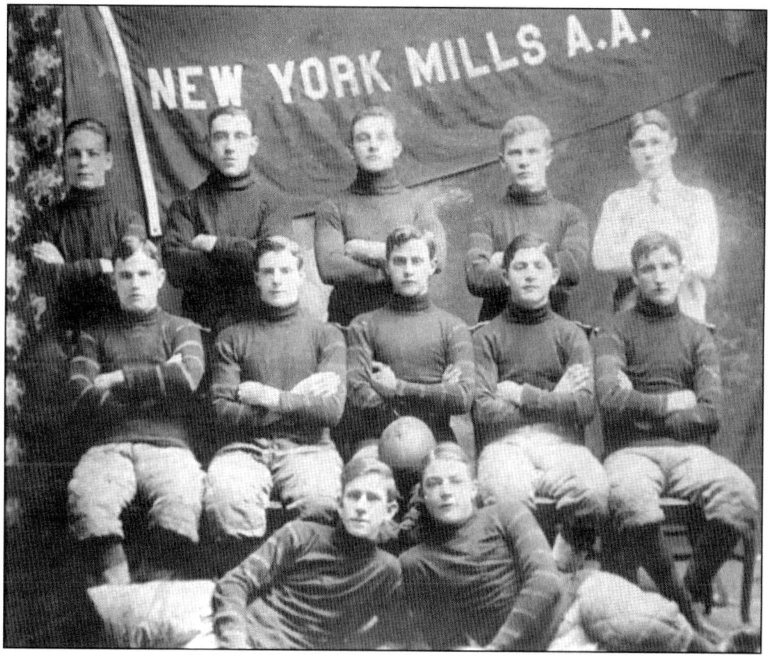

The New York Mills Athletic Association fielded this football squad whose uniforms look quite different in some respects from the Burrstone group. Note also the large pennant in the background. The village fielded several football teams usually based on location—Lower, Middle, or Upper Mills—or workplace such as the dye house or the carding rooms. (OCHS.)

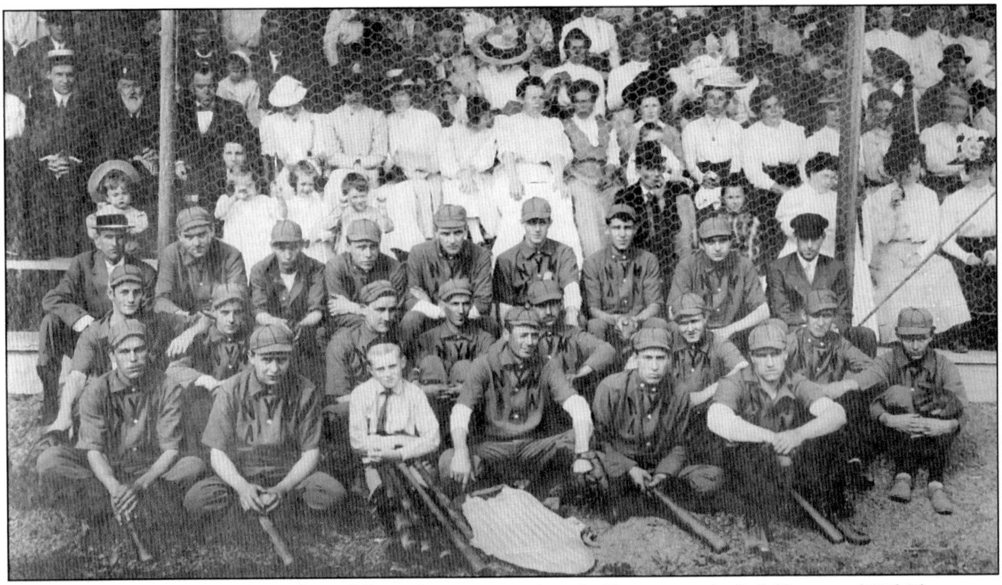

The 1910–1911 New York Mills Athletic Association baseball team included (first row) Ed Shannon (second from left), Bill Harrington (fourth from left), and Floyd Stott (fifth from left); (second row) Bobby Hughes (left) and Ed Harrington (fourth from left); (third row) manager John Shaw (left) and Hugh X. Jones (second from left). (NYMHS.)

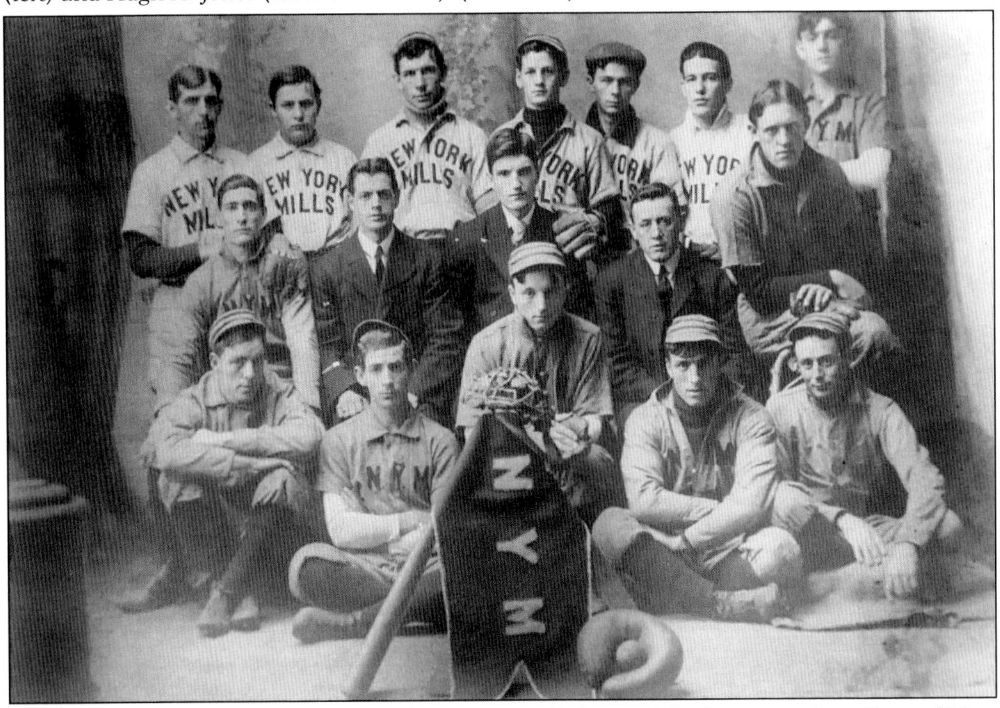

The 1903–1904 and 1904–1905 baseball team compiled one of the best records in Central New York. From left to right are (first row) Jackie Williams, Ed Harrington, Bill Jones, Bill Wilson, and Bill Harrington; (second row) Mike Mahoney, Coley Williams, Jesse Winn, Tom Stoot, and Ed Button; (third row) John Lehman, Ed Shannon, Don LaShore, Bob Hughes, George Wallace, Fred Hahn, and Hugh Jones. (OCHS.)

Three

From Labor Strife to World War

In 1908, the A.D. Juilliard Corporation of New York City took control of the New York Mills. With the change in ownership came a new directive to maximize profits, often at the expense of relations with the workforce. Company managers engaged in verbal and physical abuse, attempted to play one ethnic group of workers against another and generally treated their employees with much less respect than previously under the Walcotts and Campbells. In response, Polish workers began forming a union, while the owners resorted to firing union organizers and other intimidations. In 1911, the workers formed Local 753 of the United Textile Workers of America (UTWA) and went on strike in the following year for better treatment, wages, and union recognition. Owners hired guards who roughed up strikers, had martial law declared, and brought in the state militia to keep order. Despite this, the workers remained united and won concessions on wages and treatment. Following the strike, the company began ignoring the settlement, prompting a new strike in 1916. This time, company officials evicted strikers from their homes, hired private security guards to harass and abuse strikers, and imported strikebreakers to keep the factories running. Despite these tactics, the workers once again prevailed in a work stoppage lasting 14 weeks and one day. In the end, John Golden, president of the UTWA, called the 1916 strike settlement "one of the greatest victories ever secured."

From the fields of labor conflict, village residents next became involved in international confrontation. With the outbreak of World War I, Polish residents hoped for the reestablishment of Polish independence. However, with Poland divided between the Allied and Central Powers, relatives often found themselves on opposing sides. Several village residents rushed to join the newly created Polish army in France to fight for their homeland. When the United States entered the war in 1917, many others joined the US Army. At least 10 served in the Canadian armed forces, eight in the Polish army, and 139 in the American armed forces. Three gave their lives: William E. Brouillett, Owen Davies, and Fred Schaible.

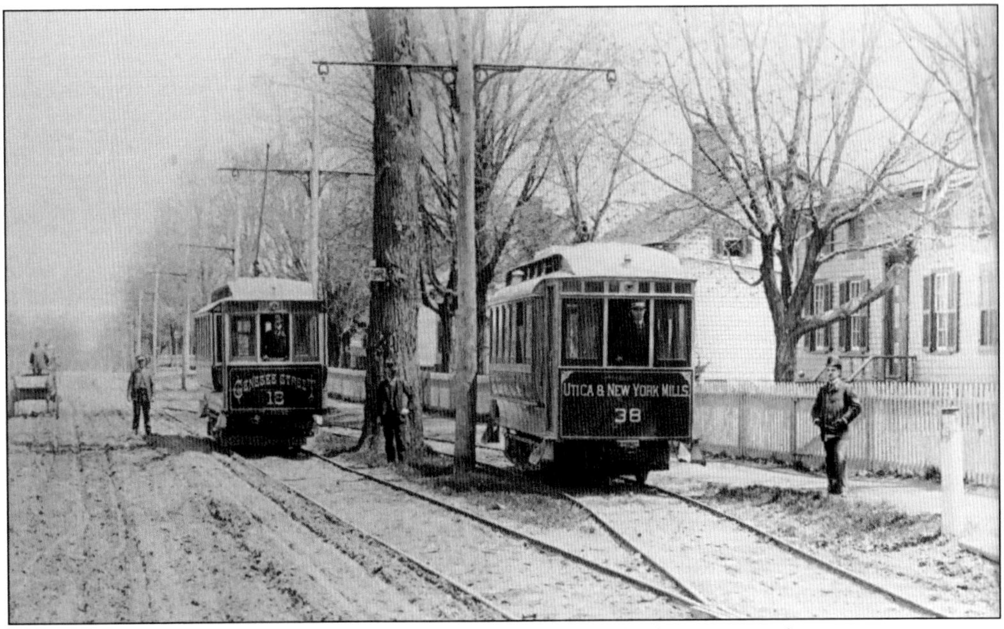

The first decades of the 20th century found the village in the beginning of a transition, as seen in this photograph where modern electric wires and trolleys share the frame with earthen streets and a horse-drawn wagon. It is believed that the small section of dual tracks to allow trolleys to pass was located in the Middle Mills area. (OCHS.)

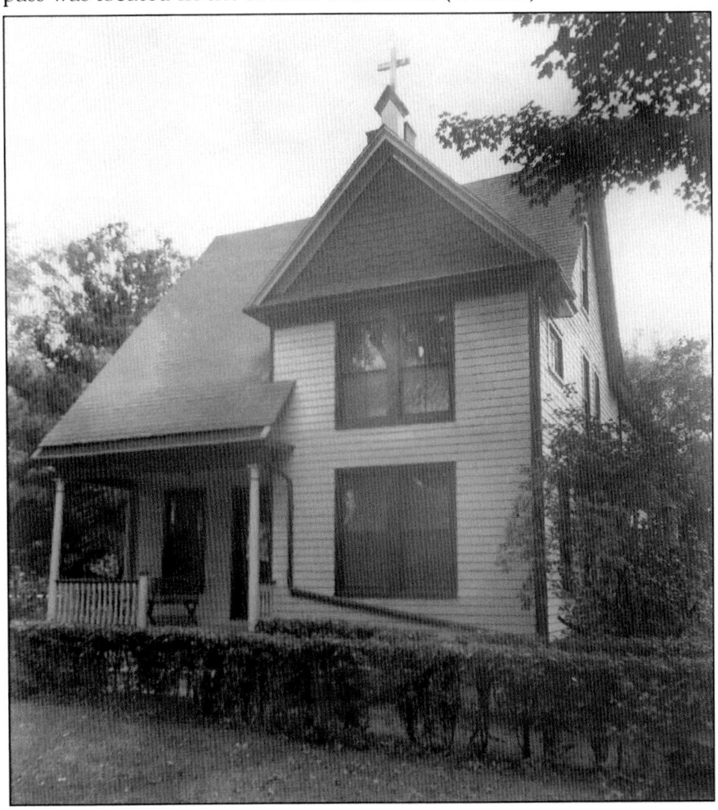

This building was originally constructed in 1884 as a caretaker's residence on the Campbell estate. In 1910, it became the convent for the Felician sisters who arrived in the village to serve as teachers in the St. Mary Our Lady of Czestochowa Parish elementary school. After two later additions, it was eventually torn down to make room for a new convent, begun with a ground-breaking ceremony on Memorial Day 1960. (NYMHS.)

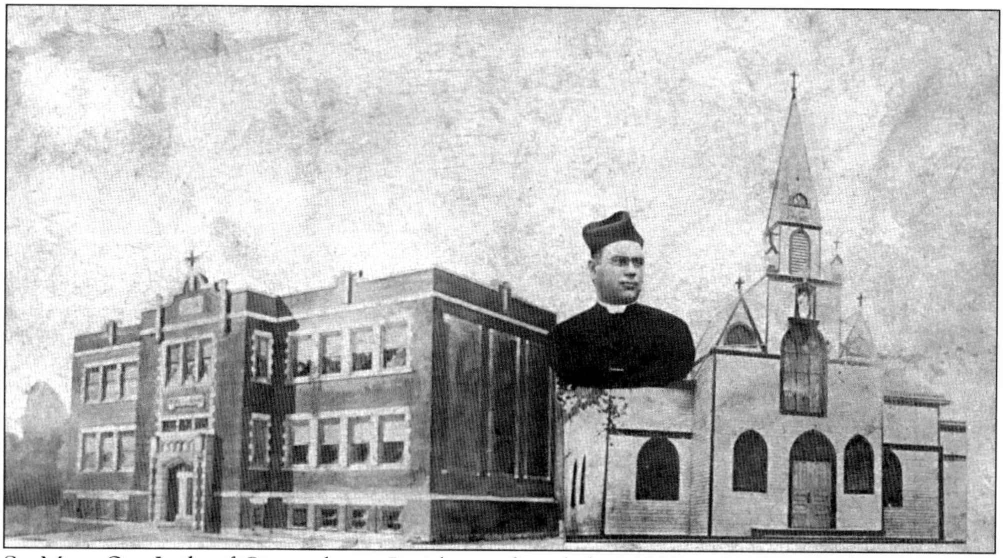

St. Mary Our Lady of Czestochowa Parish was founded in 1910 under Rev. Stefan Plaza with the first mass celebrated in the new church (right) in 1911. The school building (left) opened in 1915. The priest is Rev. Aleksander Fijalkowski, the pastor from 1911 until his death in 1958. So influential did he become in community affairs that he was often referred to as the "unofficial mayor." (Pytko.)

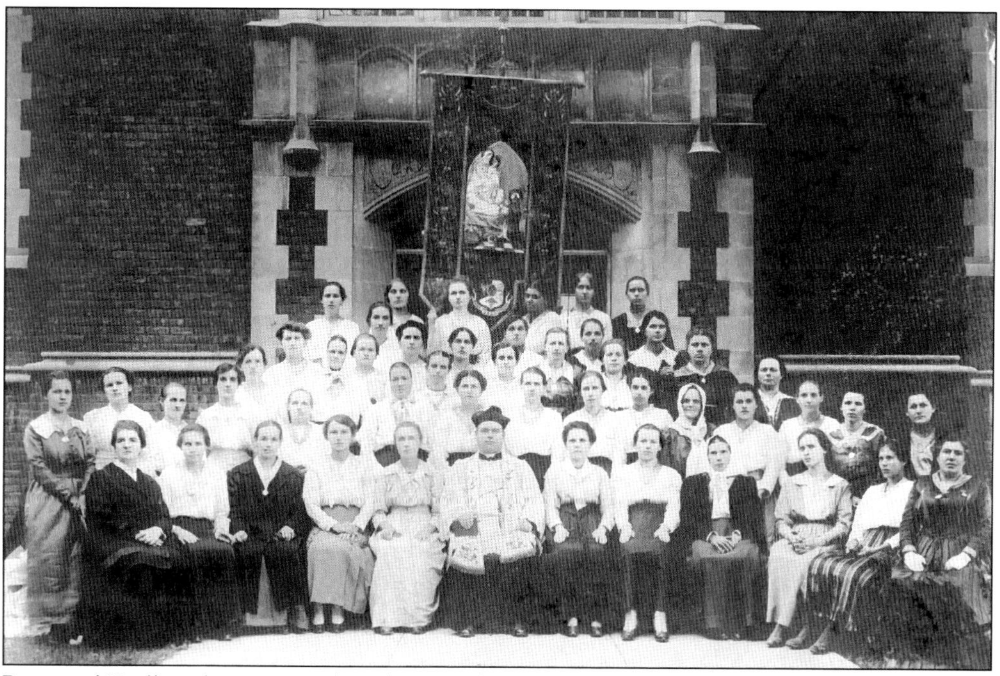

Reverend Fijalkowski poses in this photograph with one of the parish's women's organizations. Parish groups like this worked on church projects, raised money for parish necessities, and hosted social and religious events for their members. The group also provided members with an opportunity for leadership responsibilities, camaraderie, and status within the ethnic parish. (Berdell.)

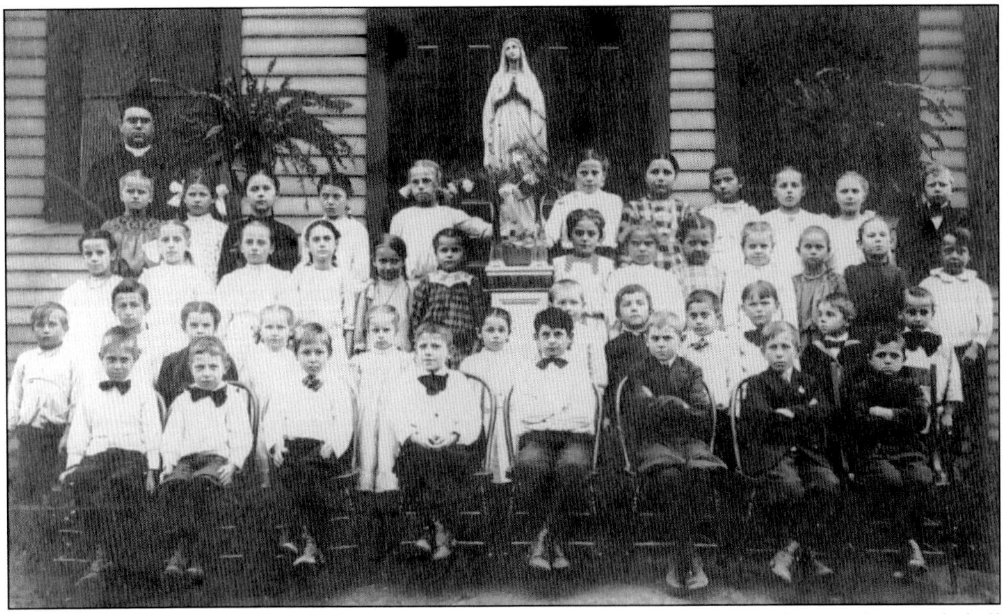

This is believed to be an early First Communion class at St. Mary's Church. The parish provided all of the religious services its Catholic community required, as well as a venue for the commemoration of particularly Polish Catholic ceremonies and national holidays for what was then a largely Polish congregation. (Pytko.)

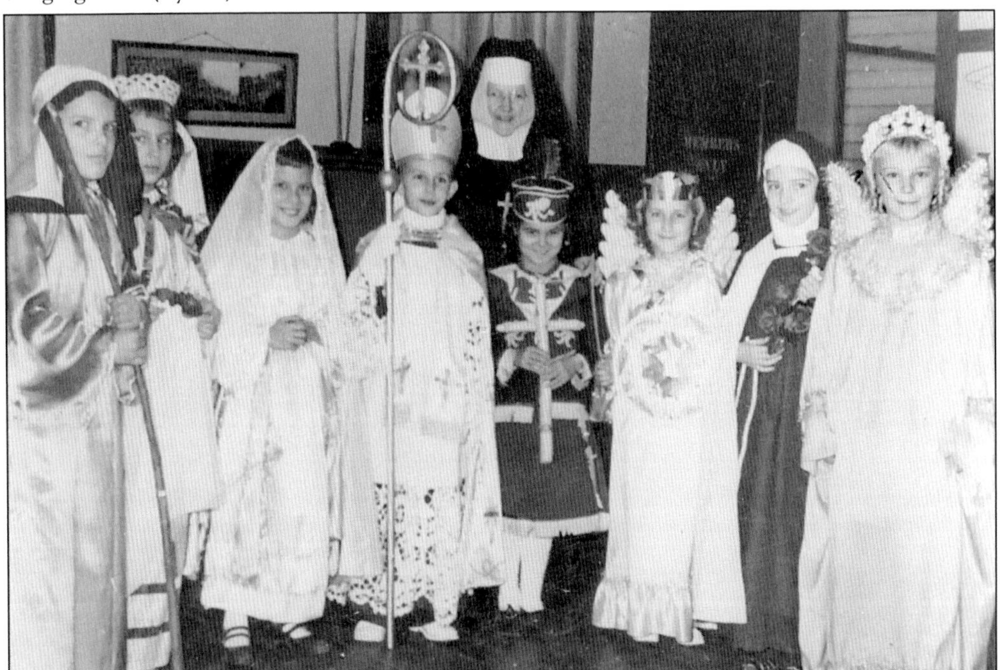

The St. Mary's parish school provided education for grades one through eight, as well as some evening classes in English and citizenship requirements for adults. In this photograph, elementary school children are staging a Christmas play under the supervising eye of one of the Felician sisters who provided the school's teaching staff. (NYMHS.)

E.M. Curtis's Shoe Shop was one of many specialized stores that operated in the village. The poster on the shutter next to Curtis urges customers to "Try Shinola, the Wonderful Shoe Polish." The wall at the left of the door contains another Shinola advertisement plus a poster announcing an upcoming dance. (NYMHS.)

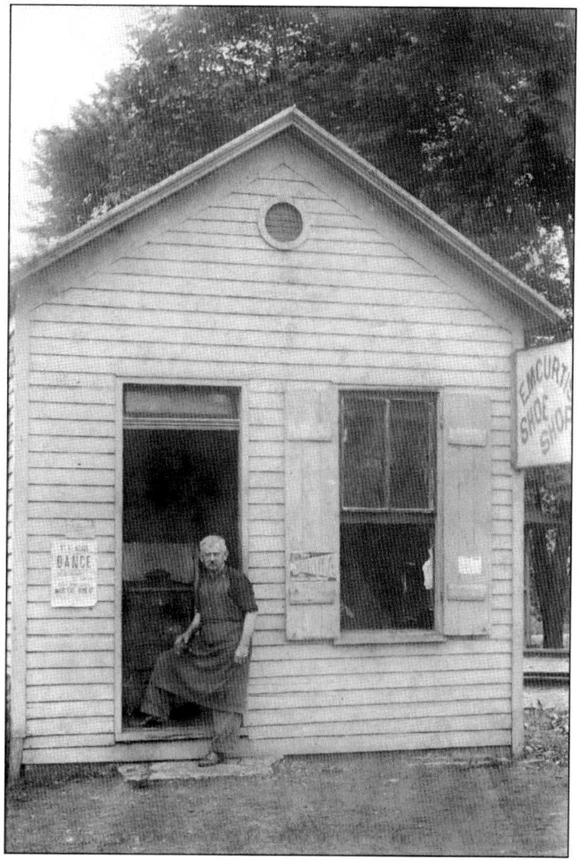

The interior of Charles Samel's store was open and functional with cutting boards and a scale for weighing purchases to the left, canned and bottled goods along the far wall, and dry goods and other merchandise to the right. Samel was a well-known supporter of church, civic, and workers' events. (NYMHS.)

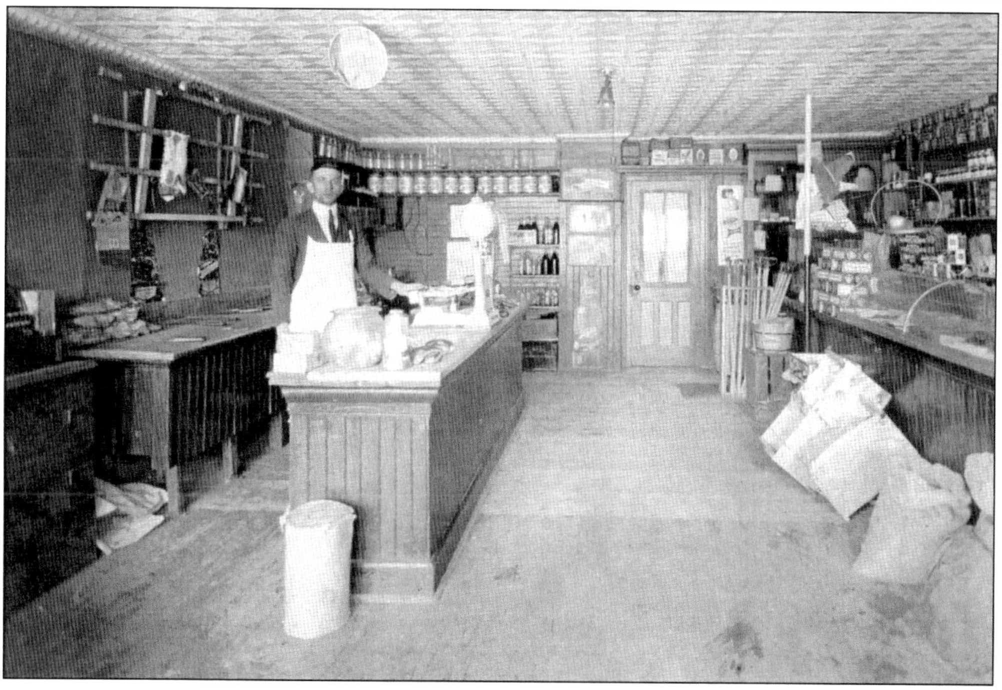

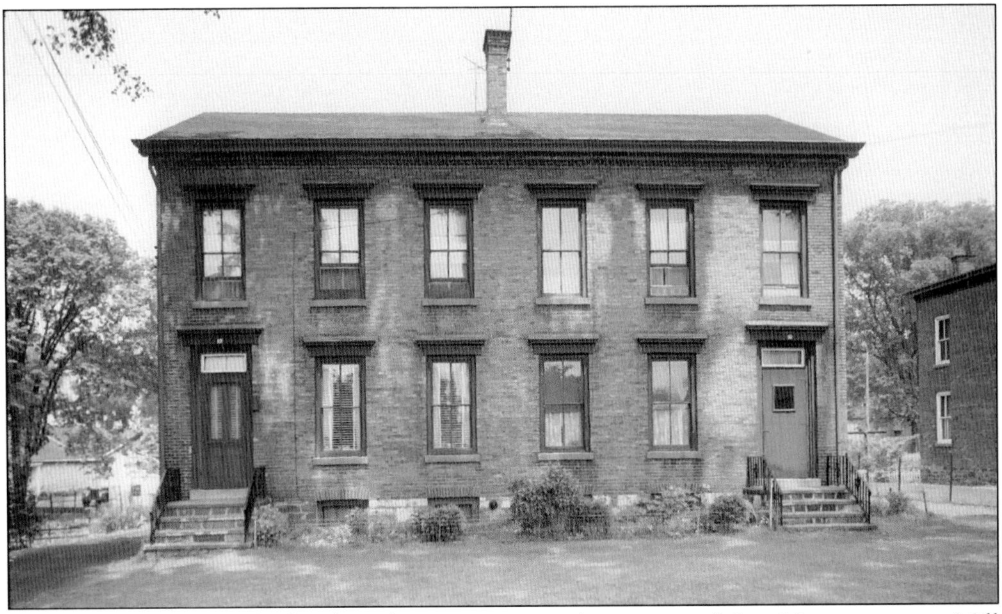

To provide more space for families, the company tore down some of the older cottages on Mill Place between 1874 and 1888, replacing them with two-story, brick structures. Each was designed to house two families. Unlike the smaller cottage design, each family had a private entrance and an attic for storage. (NYMHS.)

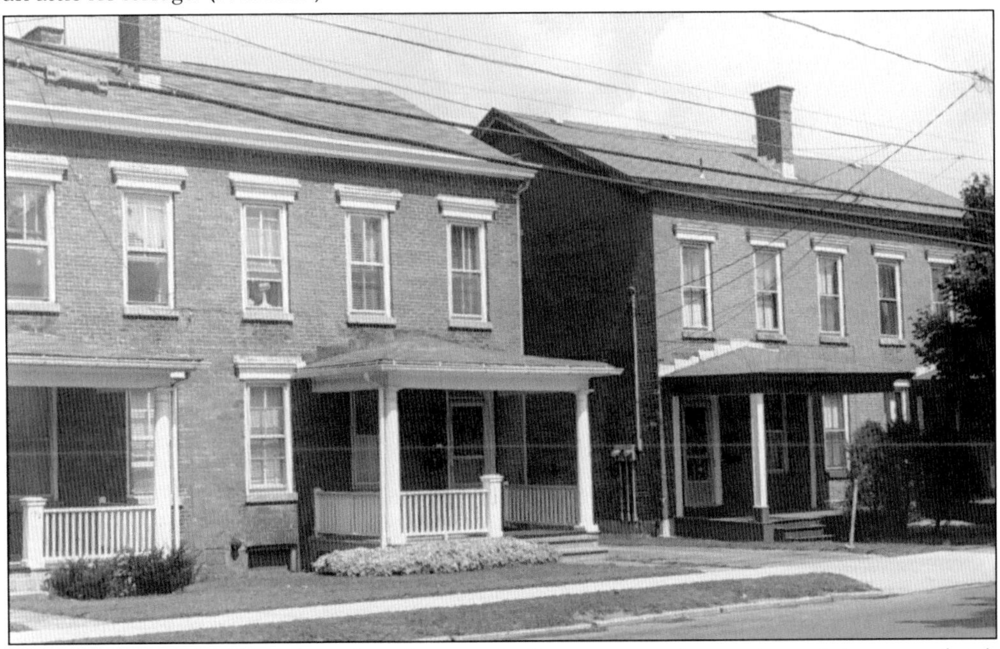

Taken at a somewhat later date, this photograph illustrates another variant of the newer brick family housing with an attached porch. Superintendents, supervisors, and company officers would normally have their own private residences. Generally, company-housing assignments were determined by the worker's position in the company. Foremen had first option on the better brick structures, skilled workers such as loom fixers second, and unskilled operators third. (Authors' collection.)

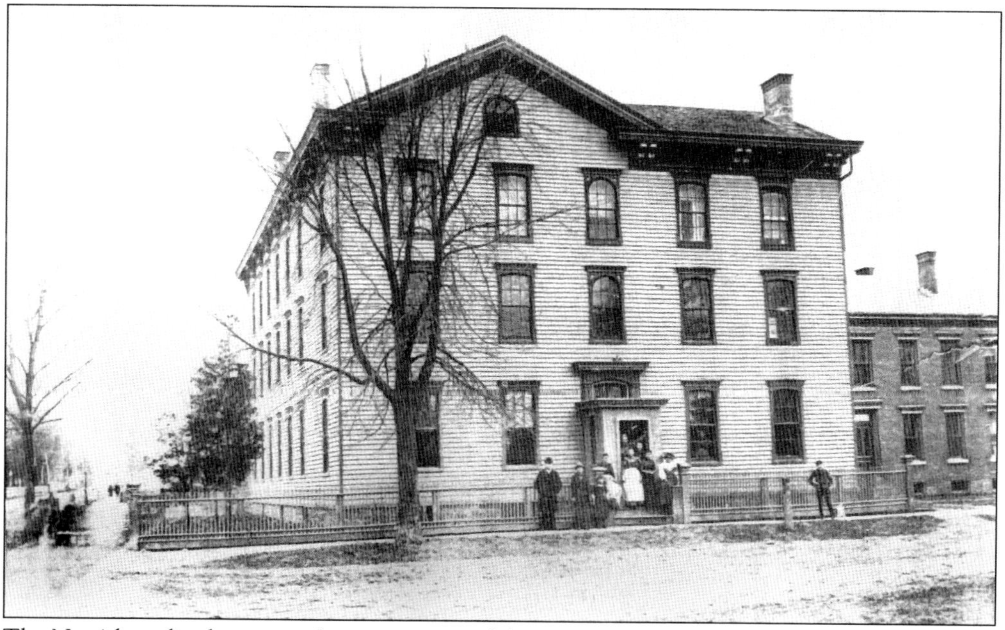

The No. 1 boardinghouse was home to many of the mill workers. Located on the corner of Main Street and Mill Place, it was only a few steps from the No. 1 mill complex. It was also a stop on the trolley line running to Utica, which brought workers to the mills from outside the immediate community. It later became Slitz's tavern and hotel. (Pytko.)

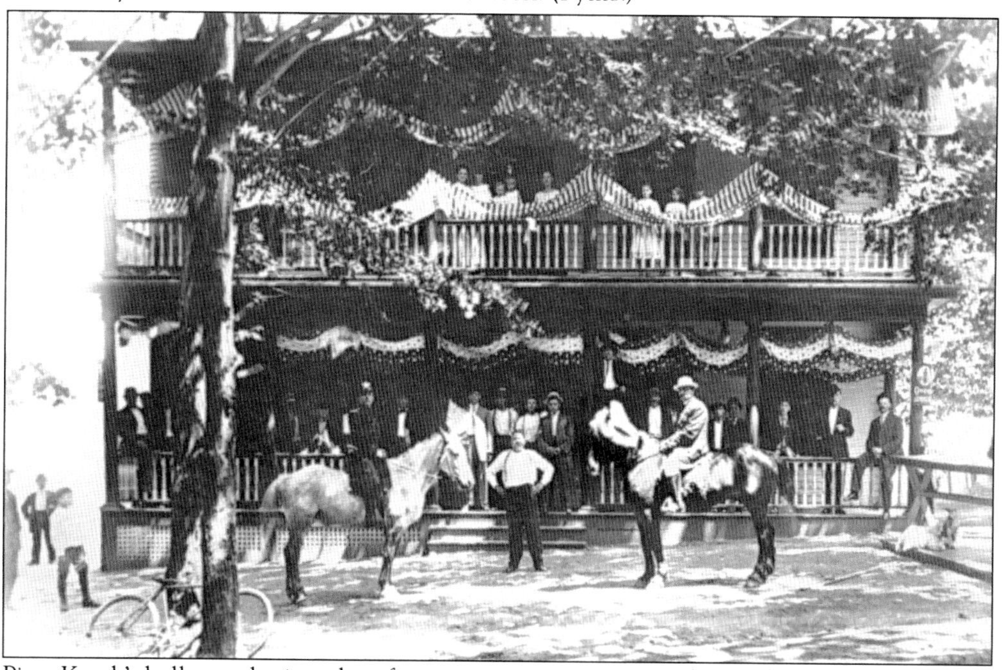

Piotr Kozak's hall, a gathering place for community events, was also used for early meetings of Polish workers planning to form a union and for gatherings of strikers once the walkouts of 1912 and 1916 began. It later became known as National Park, continuing to serve as a venue for a wide variety of community and family events well into the second half of the 20th century. (Kozak and Unczur families.)

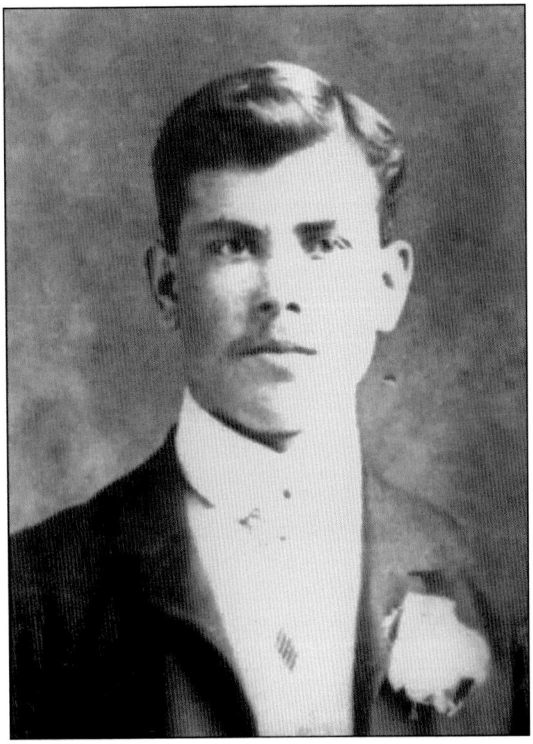

Michael Tuman was a leading organizer and first president of the union, which later became Local 753 of the United Textile Workers of America. A longtime community activist, he also supported other civic causes and served as an officer in various religious and secular societies. (Authors' collection.)

Tomasz Motyka, vice president of Local 753 and chair of its strike committee, is pictured here with a sign reading in Polish "Rownosc, Wolnosc, Braterstwo" (Equality, Freedom, Brotherhood). This became a motto of the strikers who saw themselves as fighting for equal treatment and the freedoms promised by the American Constitution. (Raymond Motyka.)

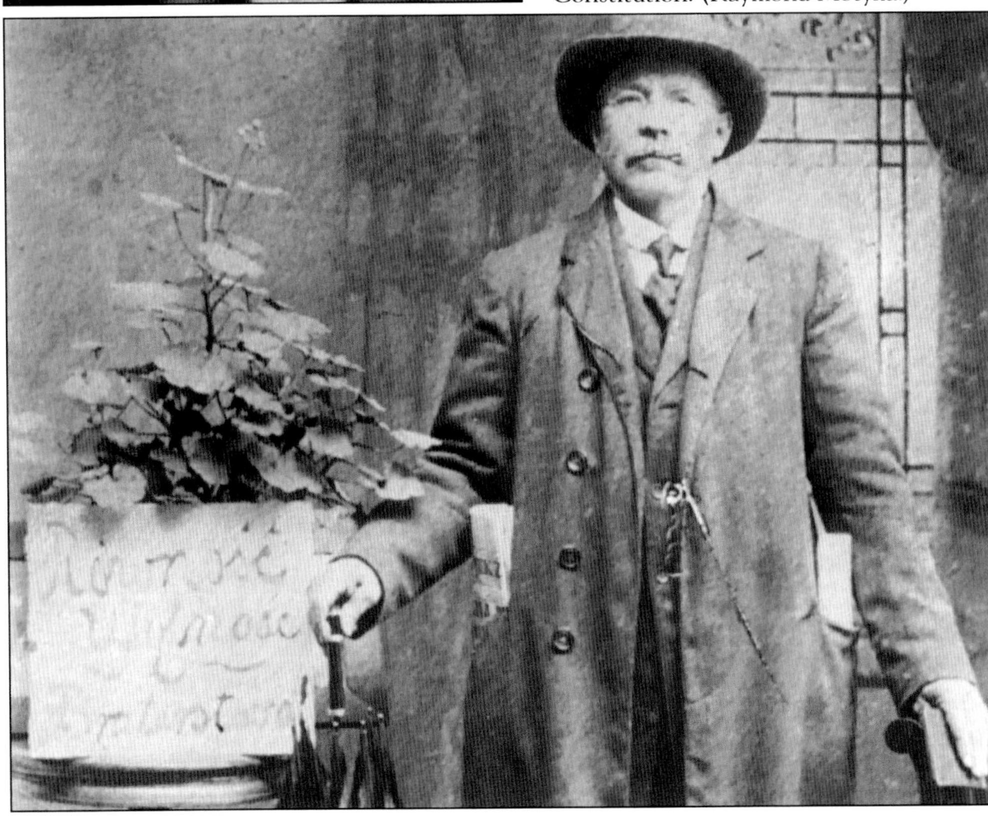

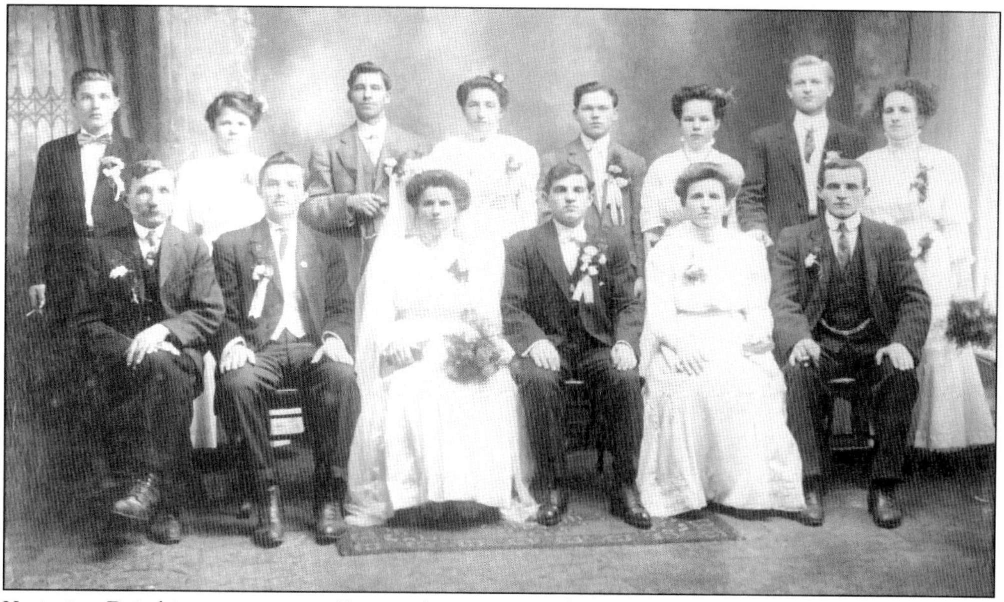

Kazimierz Dziedzic was a union officer and treasurer in the 1916 strike. Pictured in this photograph of his wedding are, from left to right, (first row) Franciszek Golab, Stanislaw Pluta, Katarzyna Golab (bride), Kazimierz Dziedzic (groom), Zofia Dziedzic, and Andrzej Dziedzic; (second row) unidentified, Katarzyna Kosmider, Jozef Tobiczyk, Rozalia Mieszczak, unidentified, Aleksandra Dziedzic, Stanislaw Pula, and Waleryja Rej. (Authors' collection.)

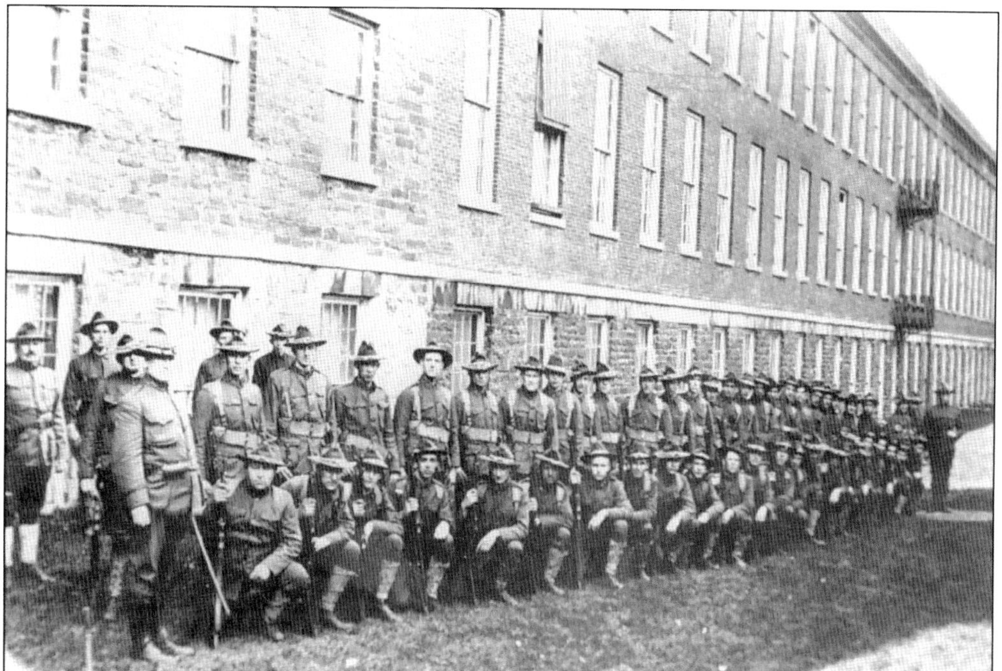

Fearing for their property and their own safely, mill owners called for help from local sheriff's deputies and then the New York State militia. Companies A, B, and M responded quickly from nearby Utica and Mohawk, placing the village under martial law and deploying to protect each of the company's properties throughout the village. (Authors' collection.)

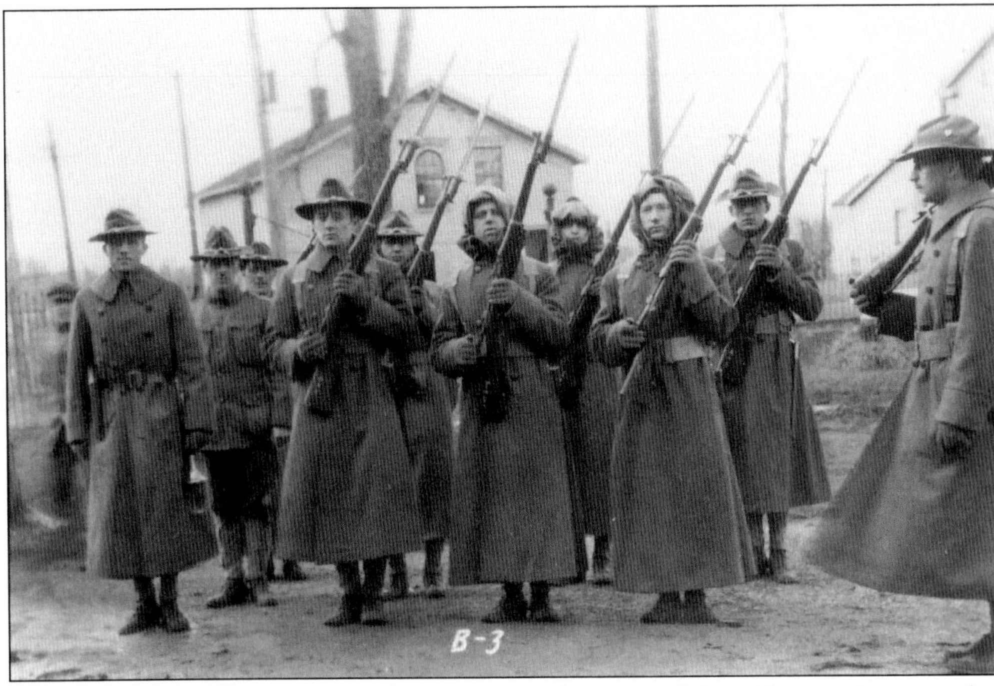

Local sheriffs' deputies arriving on the scene not only protected the mill owners' property but also often assaulted the picketers. When the militia soon began arriving, their fearsome look, as this photograph suggests, made many strikers fearful they would be subjected to more violence. Instead, the militia acted in a professional manner, preventing the deputies from attacking the workers. (Authors' collection.)

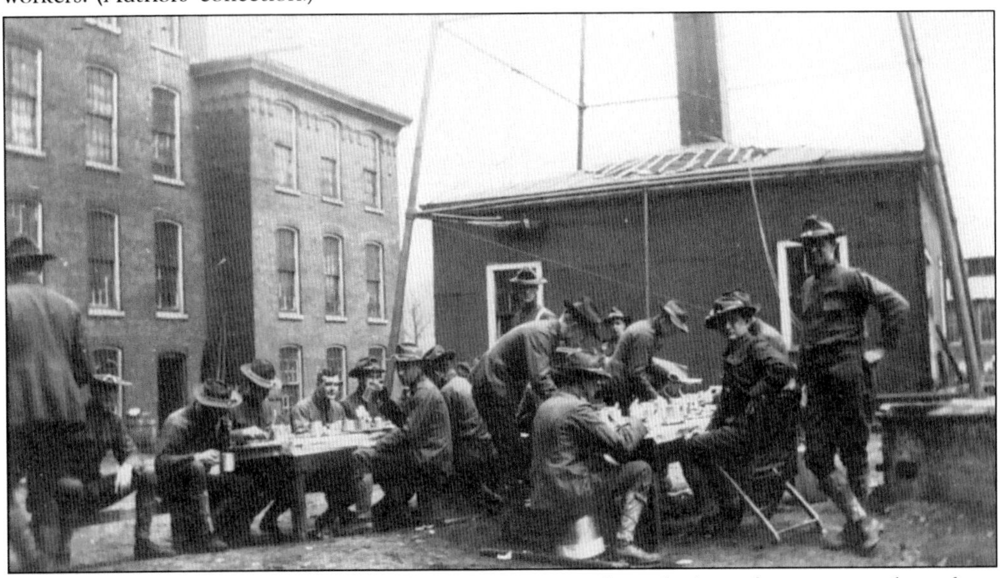

Once the strikers realized that the militia was treating them fairly and protecting them from deputies, they brought coffee and warm egg sandwiches to the soldiers, gestures that were appreciated all the more because of the cold, damp March weather. This photograph shows members of the militia at lunch. The strike gained raises and promises of better treatment from A.D. Juilliard & Company. (NYMHS.)

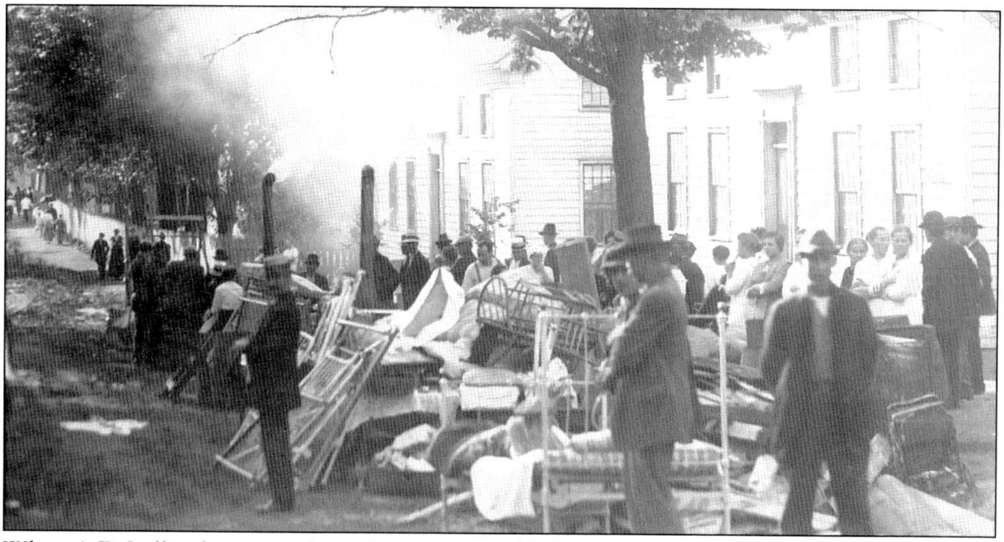

When A.D. Juilliard attempted to reduce wages and failed to abide by its agreement to treat workers better, they again went on strike in 1916. The company retaliated by evicting every striker from company-owned housing. This photograph clearly shows mattresses, bedsprings, chairs, suitcases, and even iron stoves piled by the roadside after the evictions. Once in the street, belongings were at the mercy of the weather. (Authors' collection.)

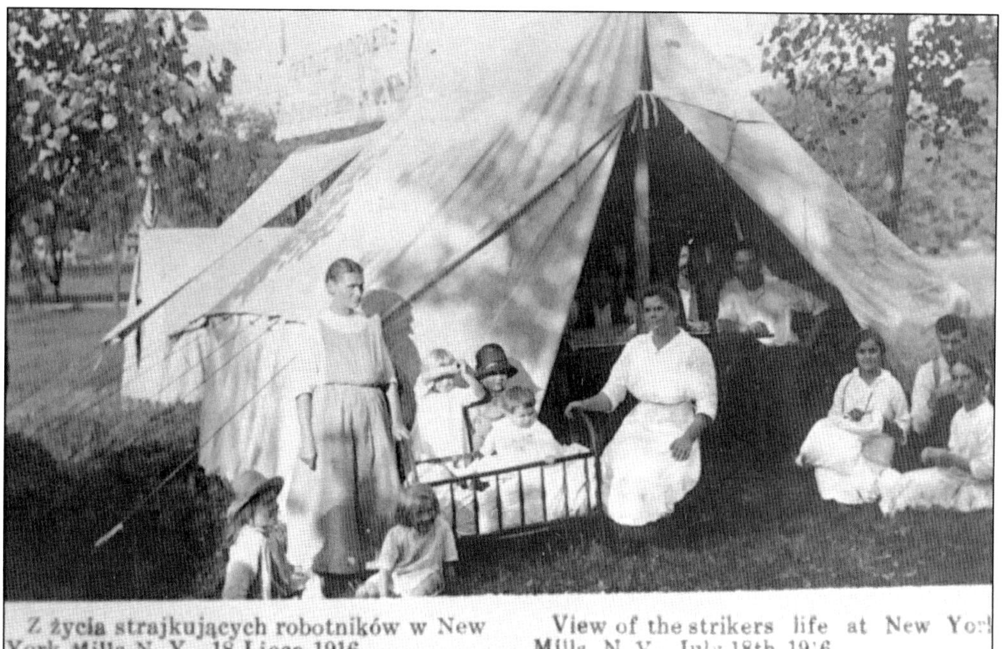

The strikers set up a tent encampment depicted in this photograph and in the cover illustration. They also hired a photographer to take pictures, which they then printed as postcards. They sold these to raise money to support themselves while out of work. Each postcard had a caption in Polish and English as shown here. (Authors' collection.)

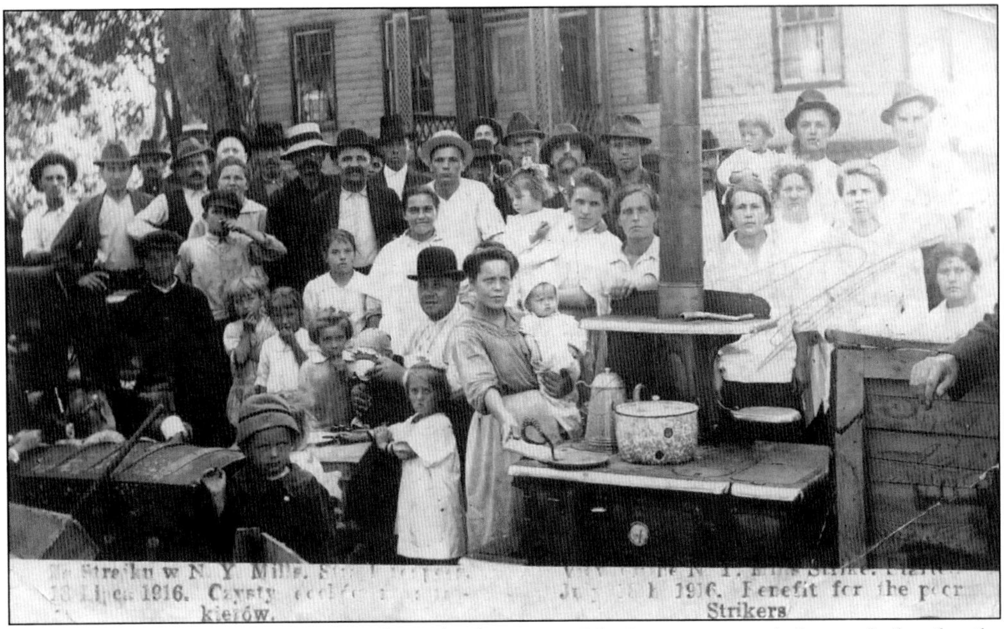

This woman, believed to be Mrs. Lorek, took advantage of the lit stove to cook dinner for her family. The little boy in the foreground is Zenon Chrabaszcz. Postcards like these helped turn public opinion in surrounding communities in favor of the strikers, leading to donations of food, tents, money, and other support from other area unions and the general public. (Authors' collection.)

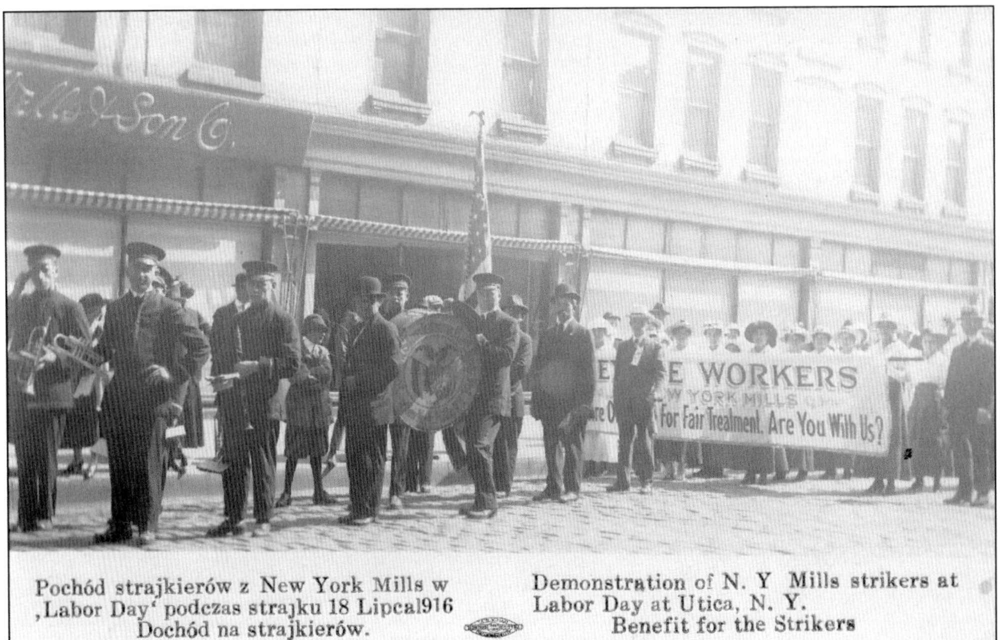

The St. Stanislaus Society band led striking workers along Genesee Street in Utica during the annual Labor Day Parade in 1916. The strikers used the occasion to draw attention to the strike and raise public awareness, sympathy, and funds to support the strikers and their families. The union proved very adept at public relations during the lengthy work stoppage. (NYMHS.)

New York Mills contained an active Syro-Lebanese population including, from left to right, brothers Ferris Nassif, Shibal Nassif, and Tony Nassif. It is believed that all three worked in the textile mills. Rose Nassif recalled wrapping thin Syrian bread around vegetables for the men to take to work. "Perhaps," she joked several decades later, "I invented the 'wrap' and did not realize it." (Rose Nassif.)

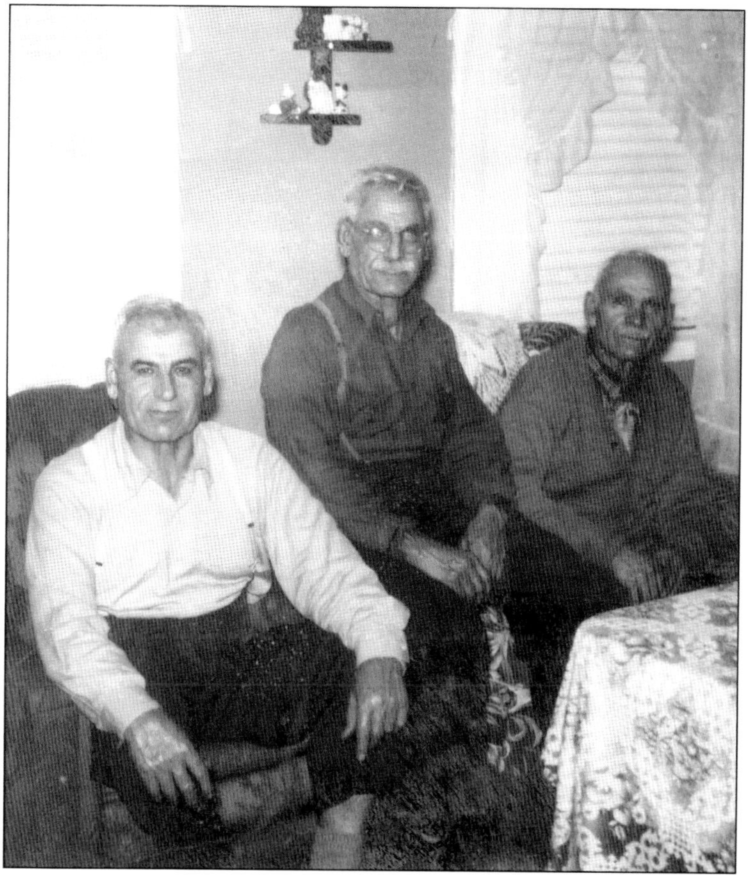

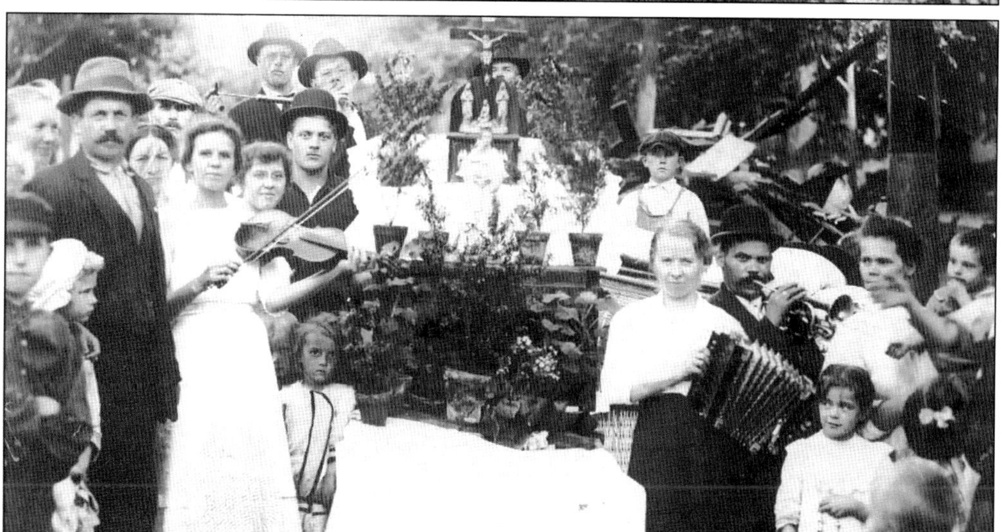

In this photograph, musicians turn the evictions into a giant block party to help the strikers pass the long hours of tedium. In the end, they won a complete victory gaining almost every one of their demands including recognition of their union, Local 753 of the United Textile Workers of America, as the bargaining agent for workers in all of the mills. (Authors' collection.)

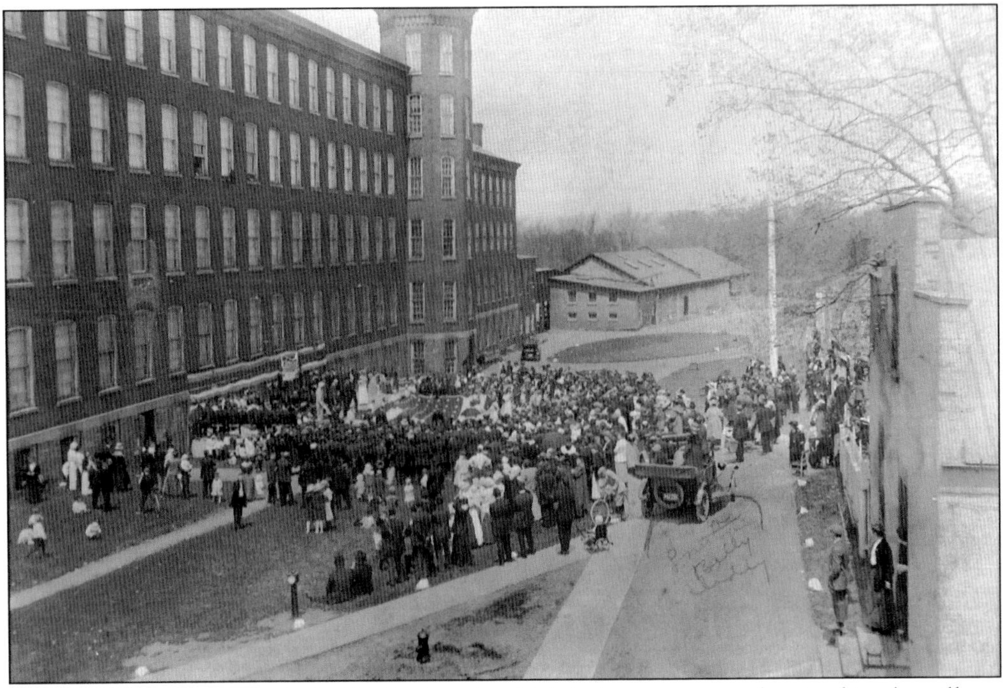

The United States entered World War I in April 1917. As soon as word arrived in the village, flags appeared, public meetings discussed ways to support the war effort, recruitment drives began, and the textile mills began a payroll deduction plan for workers who wished to purchase Liberty Bonds. This photograph was taken at a flag-raising event in front of mill No. 1 on May 19, 1917. (Montrose.)

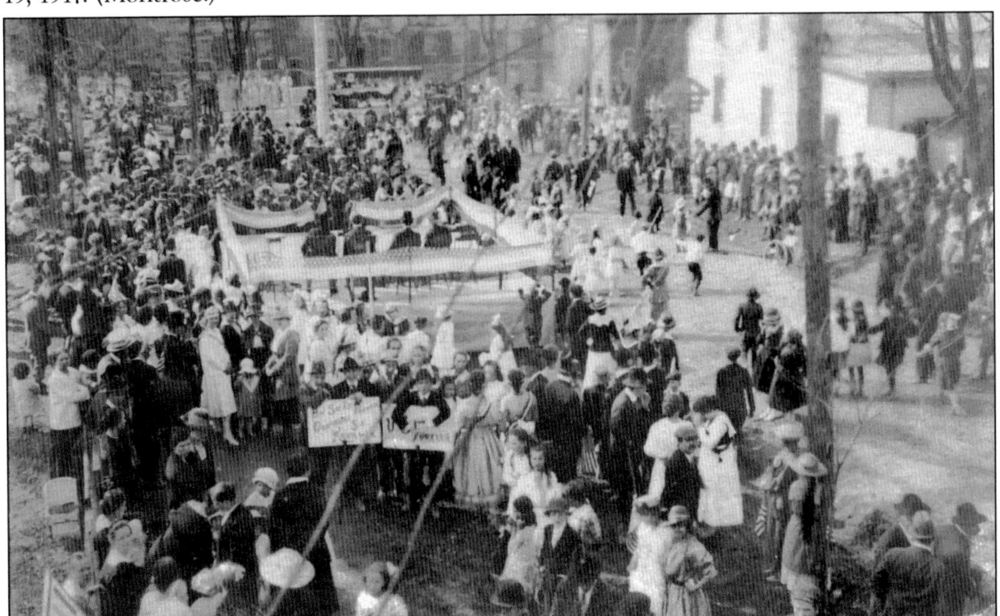

This photograph was taken at a Liberty Bond rally held in the Middle Mills in 1917. Among the dignitaries on the platform in the center was the speaker for the occasion, Prof. F. Synakowski. The distinctive Federalist facade of the No. 2 mill can be seen in the background. (Montrose.)

This truck, festooned with American flags, colorful ribbons, and patriotic young ladies, participated in one of the wartime parades in the village. Collections were taken up throughout the community to support the war effort, and special payroll deductions were arranged by the textile mills so employees could purchase Liberty Bonds through a series of weekly payments. (Montrose.)

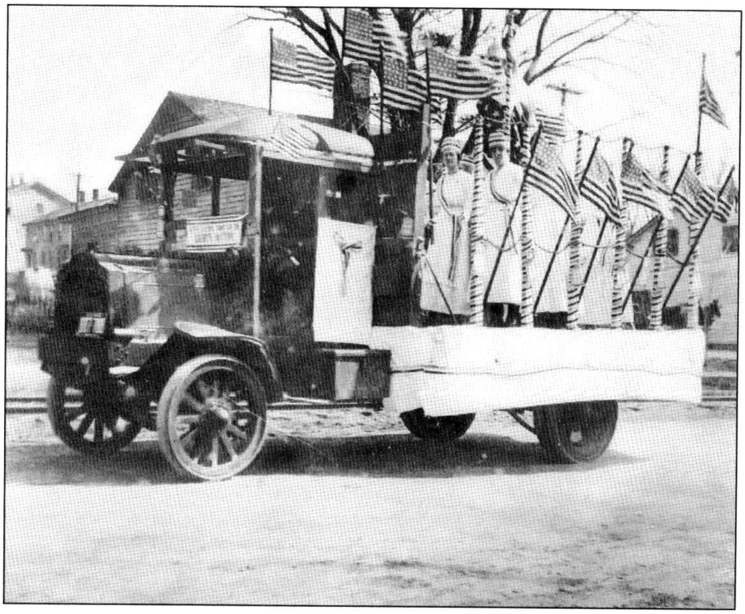

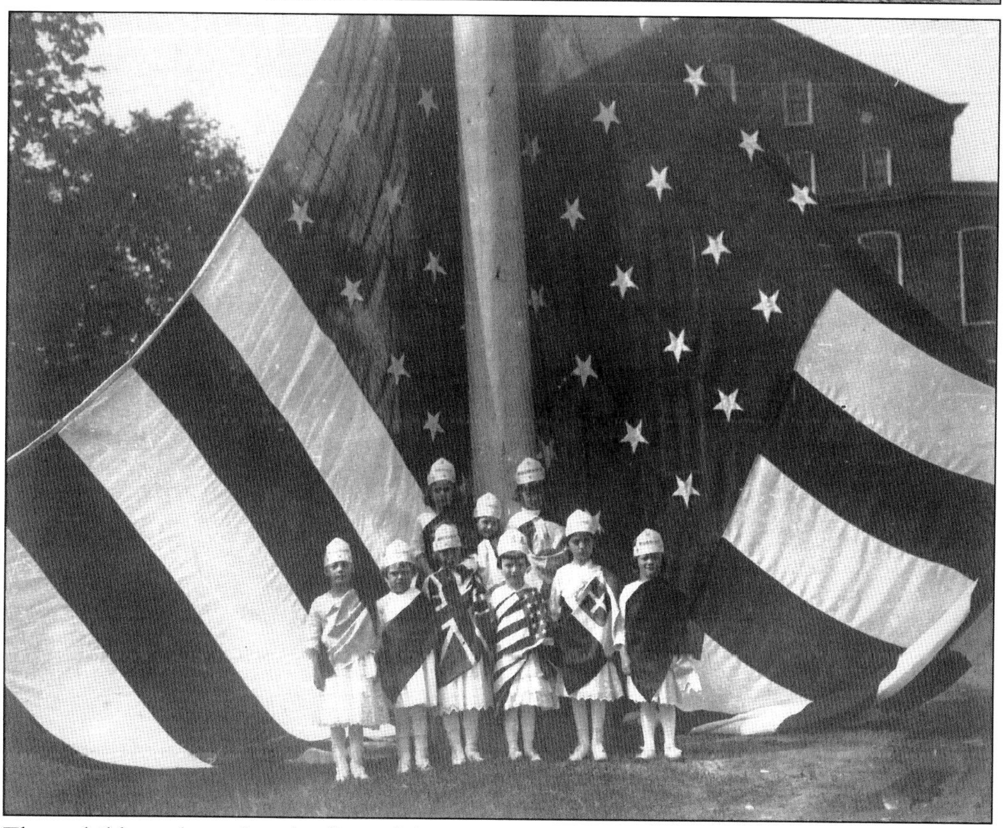

These children, dressed in the flags of the major Allied nations during World War I, are posing in front of the large flag raised near the No. 1 mill. While all segments of the community came together during the war, it was particularly difficult for those of German heritage and for Poles whose kin in Europe were often drafted into opposing armies. (Montrose.)

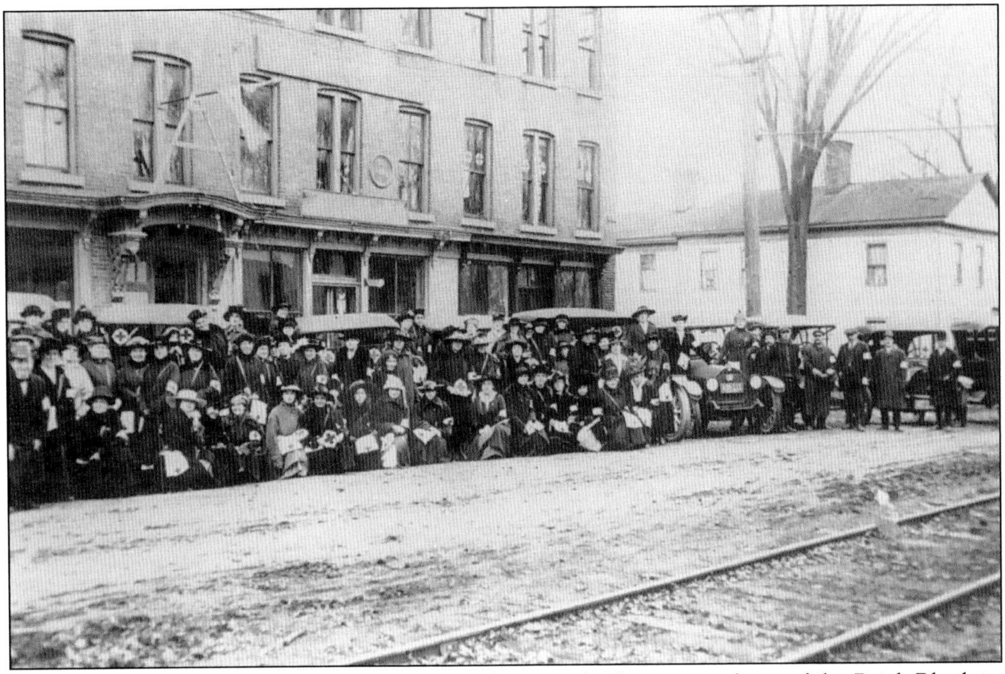

This large group of Red Cross volunteers was photographed posing in front of the Brick Block in the Middle Mills. Although the occasion is unidentified, it was probably taken during one of the numerous parades or other events staged to promote the sale of Liberty Bonds and show general support for the war effort. (OCHS.)

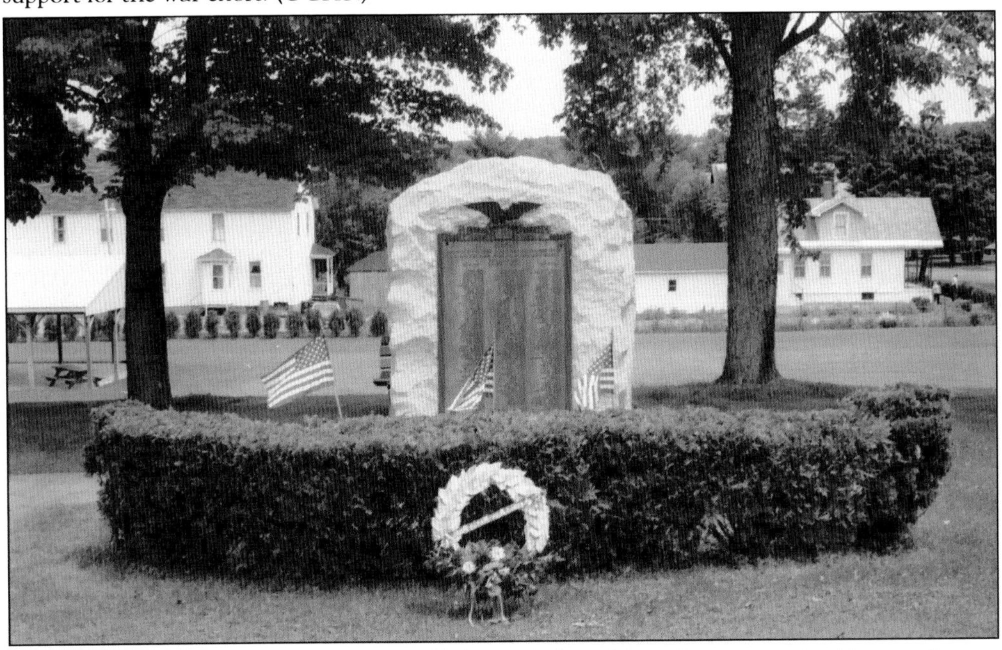

The World War I monument was originally located along Main Street opposite Pleasant Street. Following the creation of Veterans Park, it was moved there. The monument is dedicated "in honor of the boys who offered their lives in the Great War 1914–1918." It contains the names of 171 residents who served their country, including the three who lost their lives. (Authors' collection.)

Four

Prosperity and Depression

The 1920s was generally a period of prosperity for industrial America, and so it was with New York Mills. Jobs were available, wages increased, lower-priced consumer goods were more readily available, leading to a boost in the standard of living of the average worker. Small businesses flourished. New community organizations developed. At the beginning of the decade, the village underwent an important change in the evolutionary process from company town to civic municipality. On March 29, 1922, the secretary of state of New York issued a certificate of incorporation for the Village of New York Mills. The new village officers immediately determined to make much-needed improvements, such as the installation of a sewer system, garbage removal, and fire protection. Yet, the era of prosperity soon ended.

The onset of the Depression dramatically changed village life. One skilled corduroy cutter made about $50 a week in 1928, but by 1932, wages declined to only $2.50 a week. Jobs became hard to find at any wage. Meat became a rarity at tables dominated by homegrown vegetables, thin soups, potatoes, and noodles. Some residents were forced to commute long distances for jobs or to permanently move to other communities. The population fell. With the beginning of Franklin D. Roosevelt's New Deal, quite a few people took advantage of opportunities for employment in the Civilian Conservation Corps or the Works Progress Administration. These opportunities required people to temporarily leave the community, but they provided desperately needed money that could be sent home to support families. Others chose to join the armed forces as an alternative to unemployment or the uncertainties of civilian economic life.

Despite the many difficulties, village improvement continued with the establishment of a permanent volunteer fire department, a community park, and a public library. Village streets were paved, and the new sewer system was completed. Churches sponsored social events, athletic activities provided needed diversion, and local bands played for Sunday picnics at the Roselawn, or National Park, where one could enjoy a day of relaxation for a 10¢ admission. Another dime bought a hot dog.

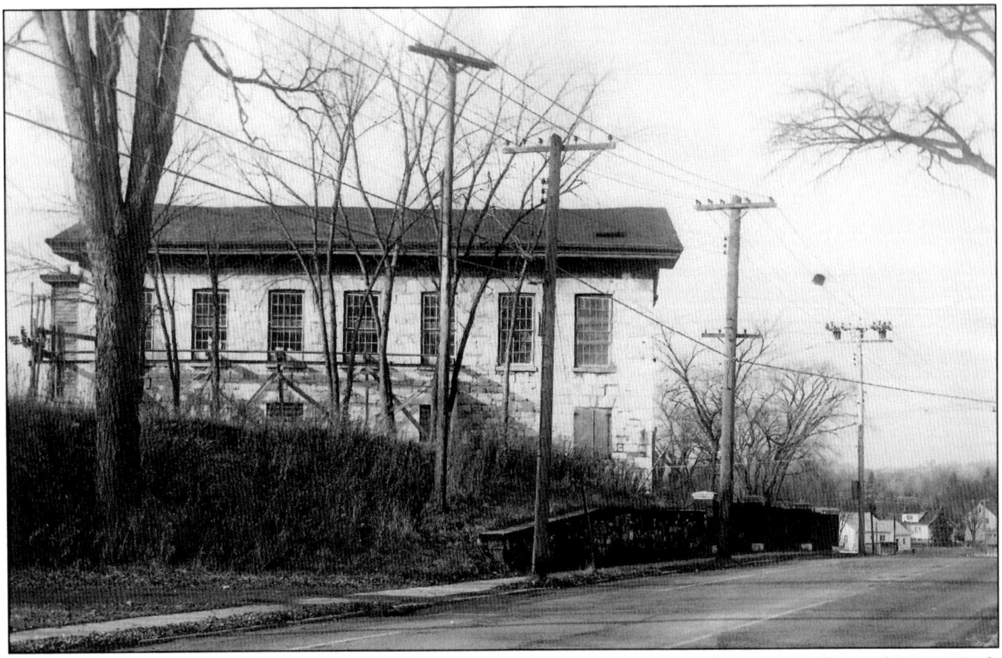

Following World War I, the national economy experienced tremendous growth, and New York Mills was no exception. The box shop in the Upper Mills was kept busy packaging and shipping goods for ever-increasing orders in the new era of prosperity. Jobs in the 1920s were plentiful, and the village was alive with activity. (Tom Stockton.)

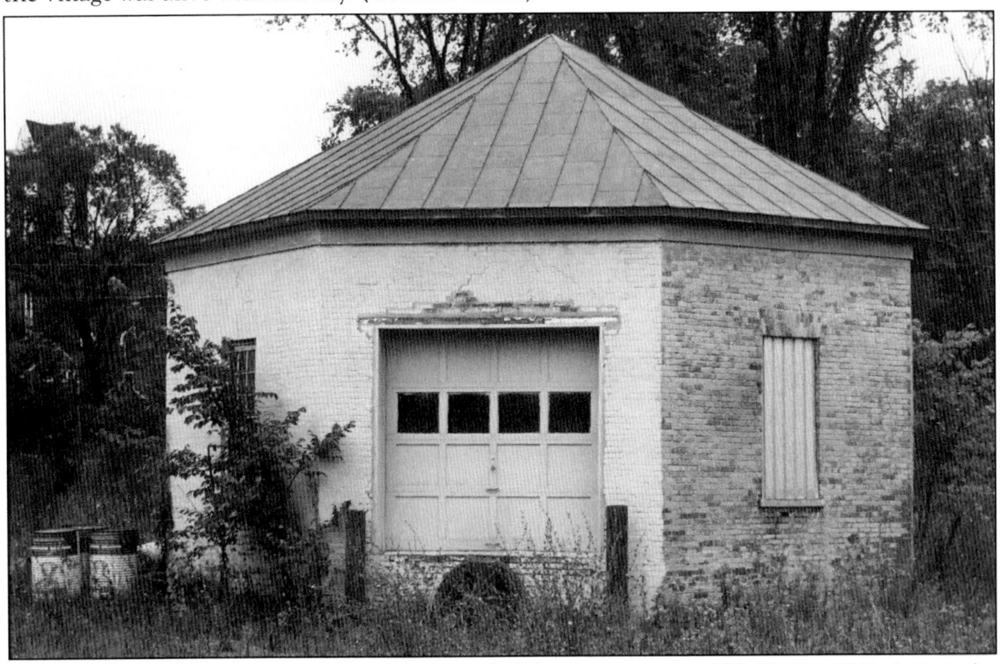

The gashouse and gasometer were constructed of brick at a cost of $2,000 in 1847 during the closing years of the period of Federalist architecture. Some four score years later, they were still in use although not always for their original purpose. The gasometer, pictured above, was largely used for storage by that time. (OCHS.)

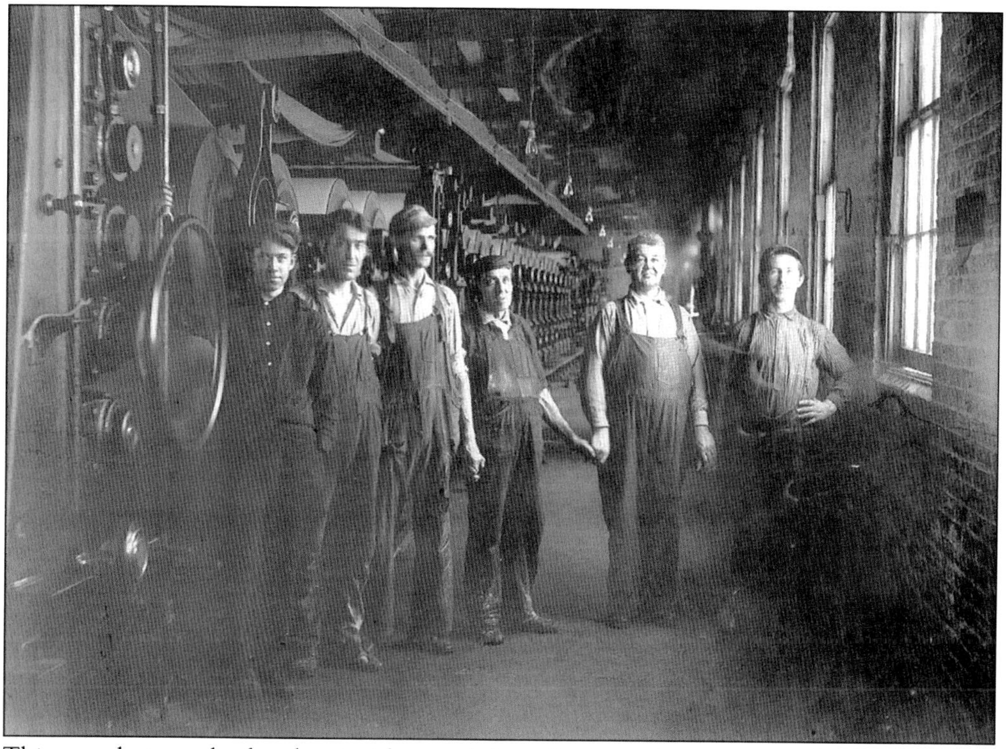

This rare photograph of workers inside one of the mills shows them posing in front of large rolls of cloth. The man on the right is John Kogut, and next to him is Hugh X. Jones. The photograph illustrates the type of overalls and other clothing typically worn by the workers. (NYMHS.)

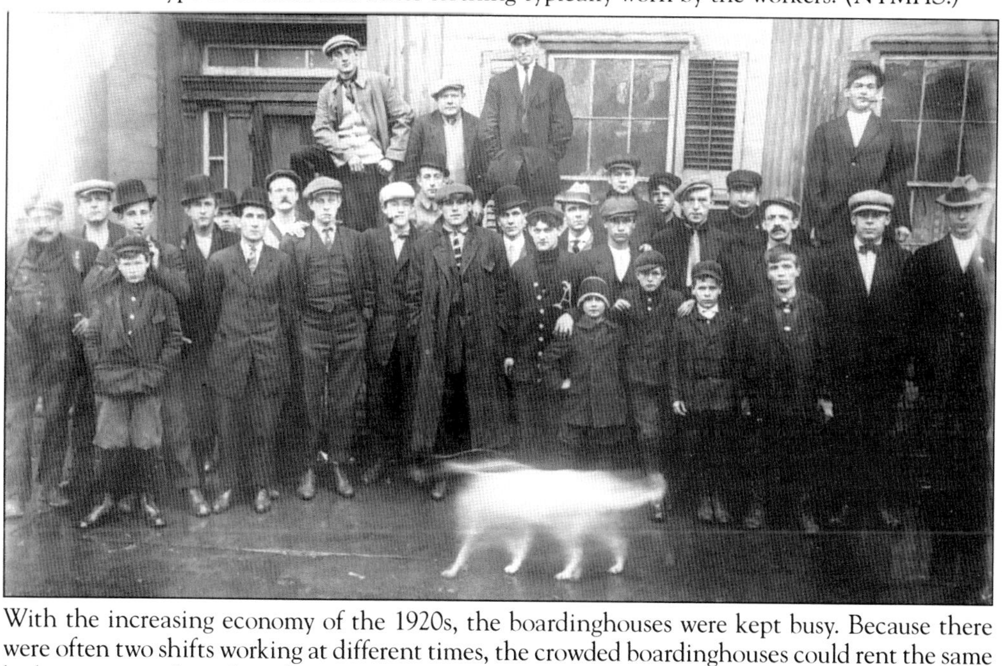

With the increasing economy of the 1920s, the boardinghouses were kept busy. Because there were often two shifts working at different times, the crowded boardinghouses could rent the same bed to two people with each using it in turn as long as they were employed on different shifts. (NYMHS.)

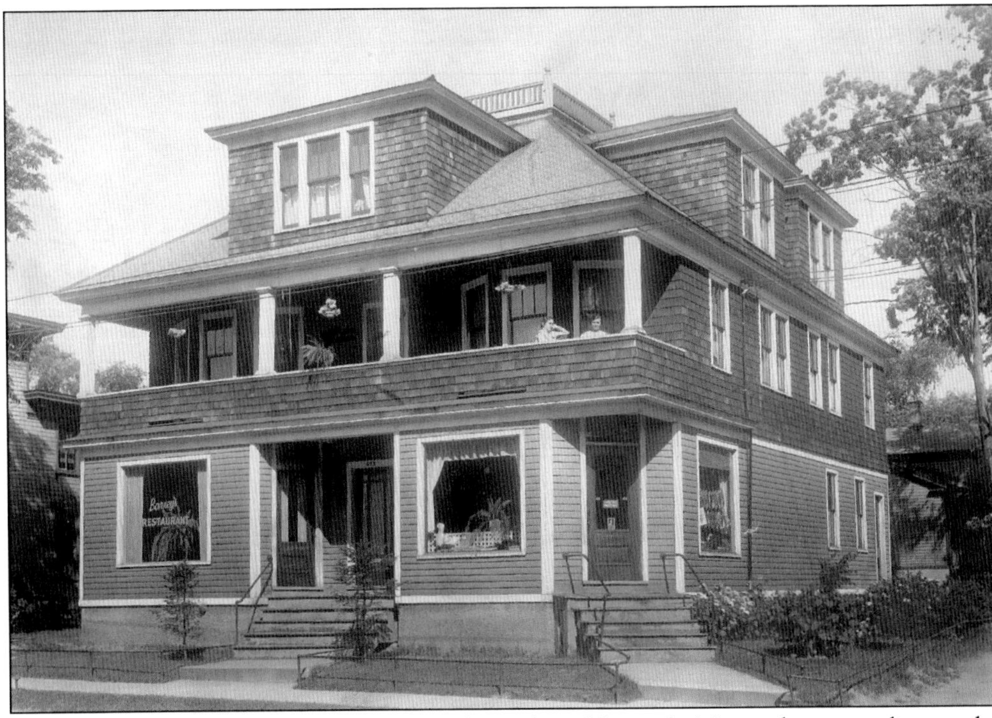

Barney's Tavern was owned by Bronislaw Klimczak and his wife, Mary, who is standing to the right of Barney's sister Zofia on the second floor. During Prohibition, the room on the right side of the tavern was converted into a soda fountain. Their daughter married Stanley Walewski, a prominent local political figure who served as Village Clerk for 27 years and used the tavern as a political headquarters. (Synakowski.)

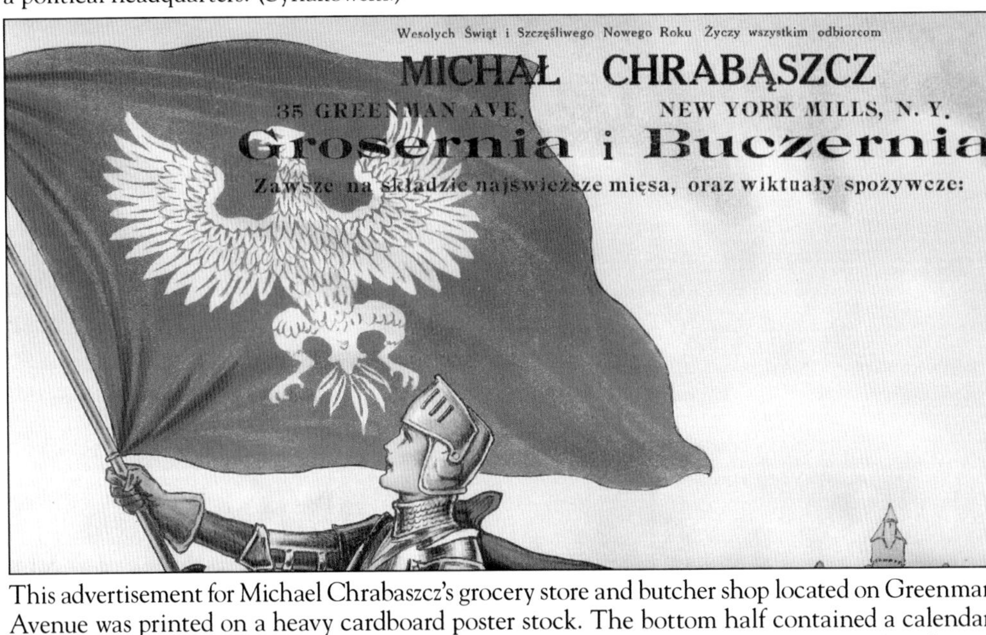

This advertisement for Michael Chrabaszcz's grocery store and butcher shop located on Greenman Avenue was printed on a heavy cardboard poster stock. The bottom half contained a calendar. During the interwar period, many businesses adopted this practice of advertising by distributing colorful calendars that people would keep in their homes. (Pytko.)

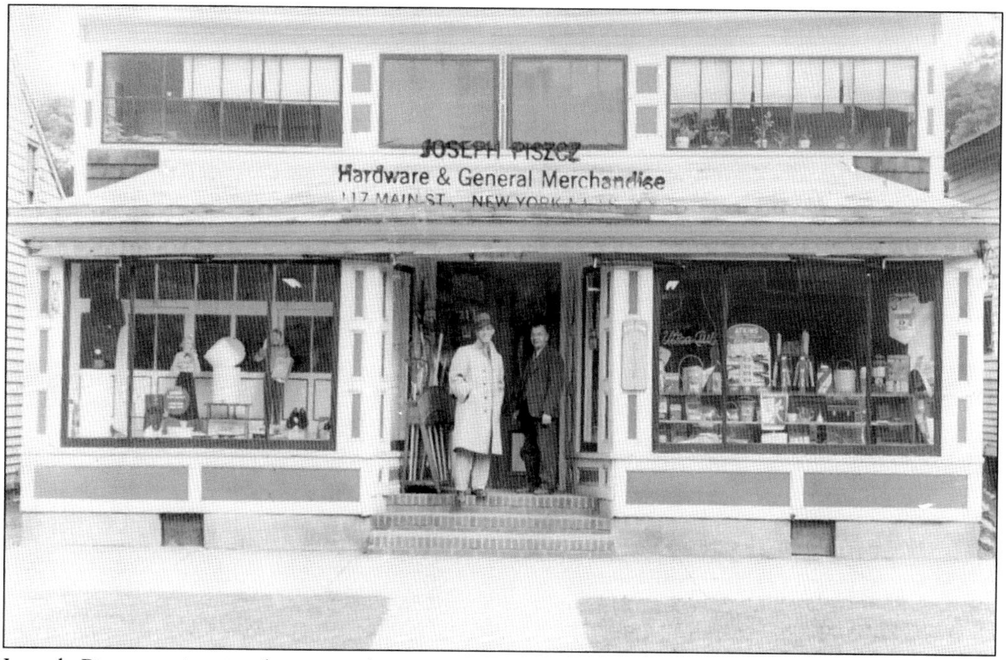

Joseph Piszcz maintained a general store at 117 Main Street in the Lower Mills. His business also specialized in hardware and some appliances. The neat appearance of the sidewalks dates this photograph to the late interwar period after the introduction of this civic innovation in the 1920s and 1930s. (NYMHS.)

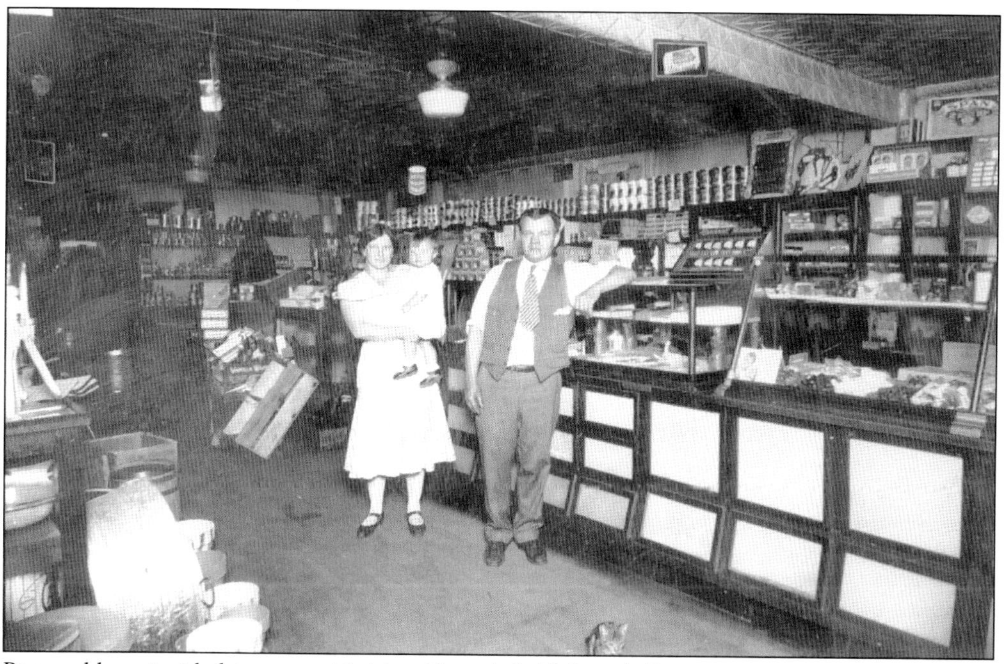

Pictured here inside his store with his wife and child, Joseph Piszcz was a veteran of the Polish army that served in France in World War I and a civic-minded individual who was a prominent member of many organizations. In 1930, he was elected a committeeman of the Whitestown Second District. (NYMHS.)

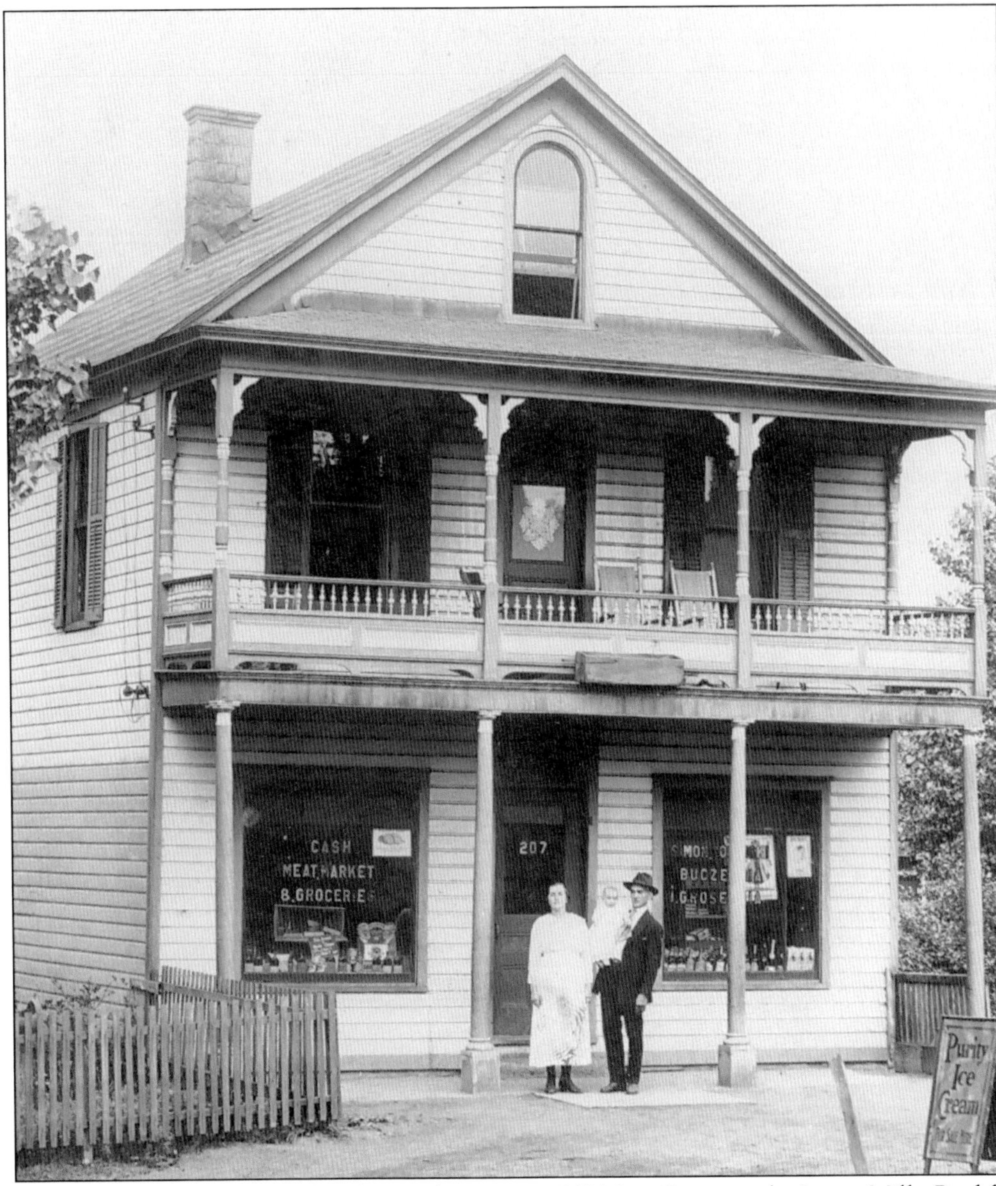

John Pezdek owned this butcher shop and grocery at 207 Main Street in the Lower Mills. Pezdek is shown here with his wife and child. Note the sign in the lower right advertising Purity Ice Cream. The writing on the windows is in English to the left and Polish to the right to attract customers from both major elements of the village's population. Many middle-class merchants owned similar establishments with their business on the first floor and living space on the second floor. This two-story clapboard construction also contained an attic and second-floor balcony complete with rocking chairs. The Pezdeks were prominent entrepreneurs, with Martin owning a grocery and butcher shop on Prospect Street and Thomas running the popular White Eagle Cafeteria at 435 Main Street in the Upper Mills. (NYMHS.)

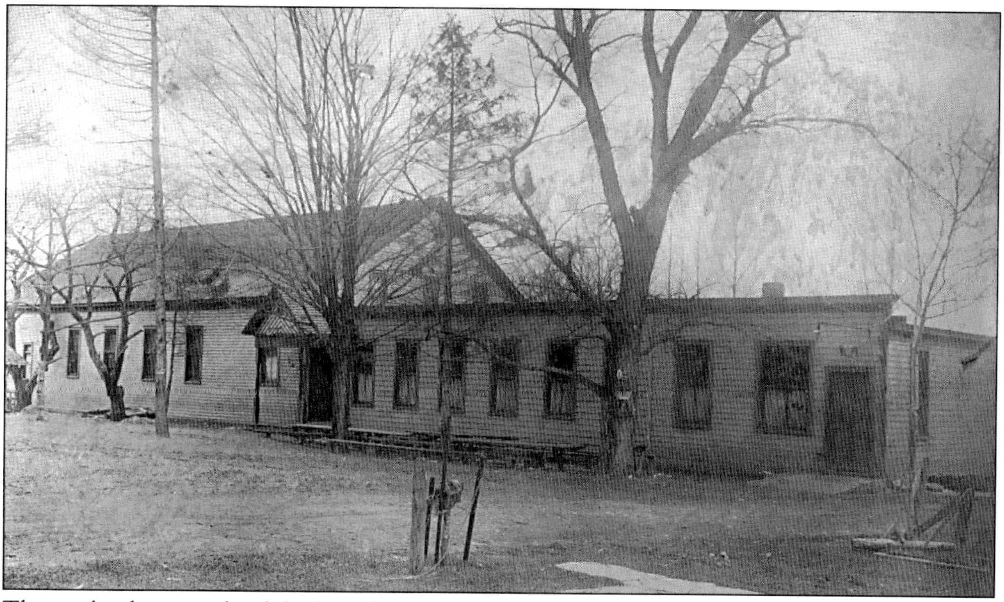

This early photograph of the Roselawn, owned by Tomasz Osika, shows the original building and first addition. Beginning as a tavern in the Middle Mills, it grew to include an expanded restaurant and later facilities for large group events such as wedding receptions and parties. Many village residents have fond memories for stopping by the Roselawn for a pizza with friends after high school athletic events. (Pytko.)

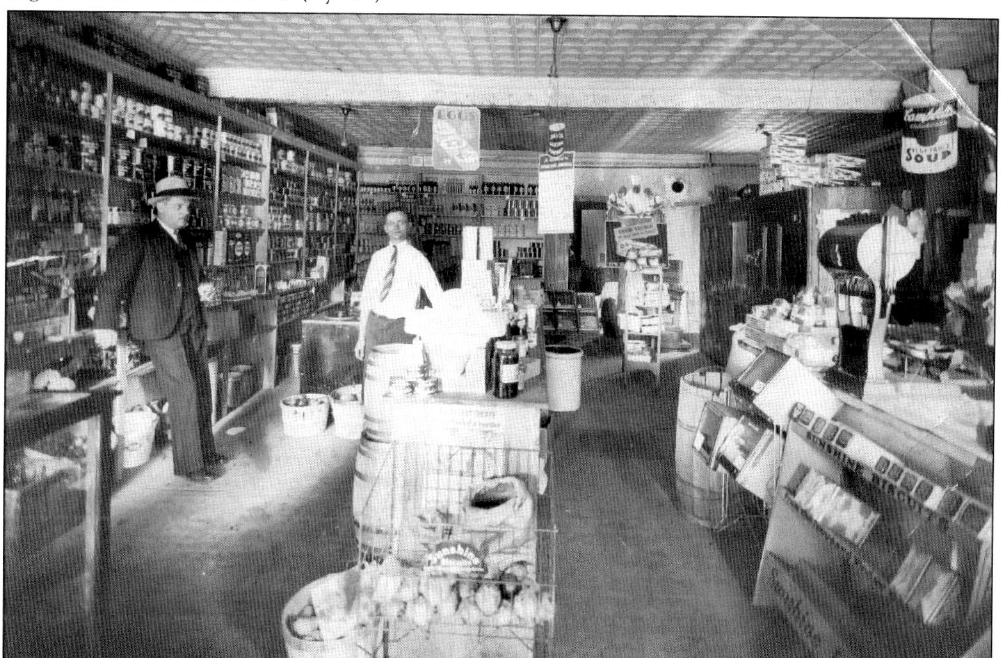

The interior of the Zawislaw Grocery Store shows well-stocked shelves of packaged and canned goods on the walls to the left and rear. In the front on the right is a scale and meat counter, while the right rear appears to have what looks like old-style ice coolers. Note also on the right the hanging Campbell's Soup sign and the tin ceiling. (NYMHS.)

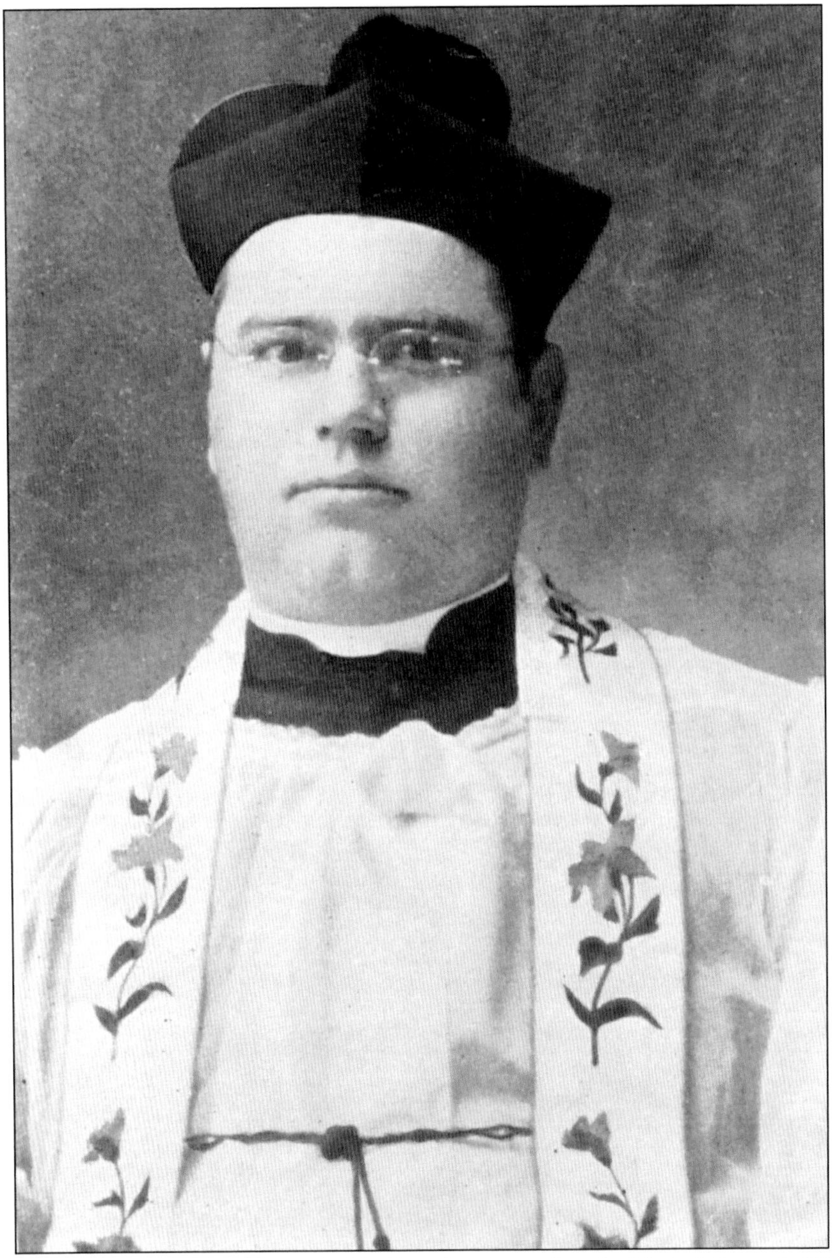

Born in Dobrzyn, Poland, Rev. Alexander Fijalkowski migrated to the United States in 1903 at age 18. After completing his religious studies at Sts. Cyril and Methodius Seminary in Detroit, Michigan, he pursued further studies for the priesthood at St. Bernard's Seminary in Rochester, New York. He arrived in New York Mills in 1911, quickly gaining a reputation for his fire-and-brimstone sermons. He served St. Mary's Parish for over four decades until his death in 1958. During that time, he proved to be a stalwart supporter of parishioners and the community, going so far as to announce at Sunday mass during the early textile strikes that parishioners should not donate to the church until the strikes were settled because they needed the money for their families. Widely respected in the community, he preached *jednosc* (unity) and often acted as a mediator between workers and management and between parishioners and government offices. (NYMHS.)

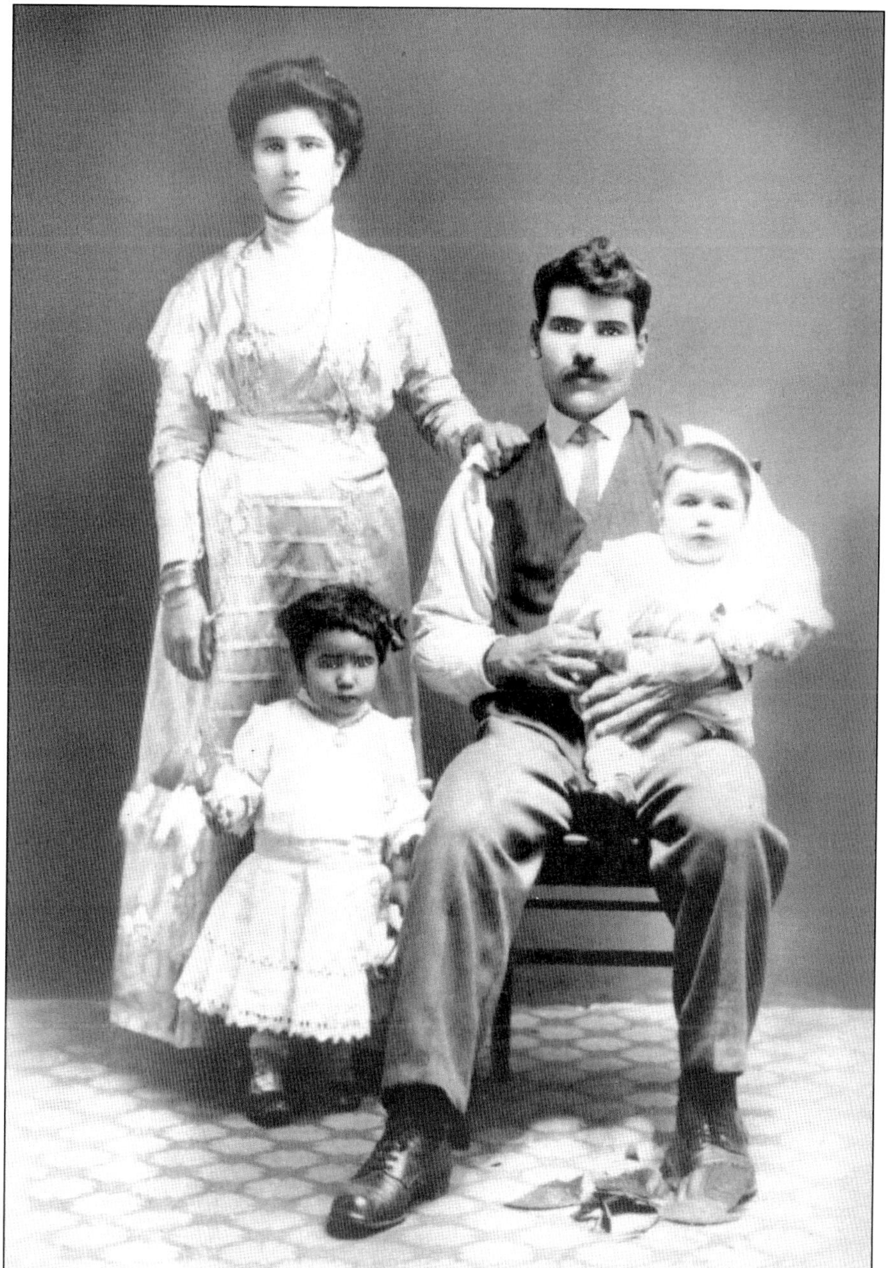

Gabriel Moses (Jibrail Aboumussi) came to America from the village of Sareen in the Bekaa Valley of Lebanon. He made a living as a peddler until he earned enough to open his own store, a typical beginning for many immigrant entrepreneurs. He is pictured here with his wife, Najla Chamoun Moses, daughter Victoria, and son John. In a classical case of chain migration, all of New York Mills's Lebanese community came from the Sareen area. Fifteen were listed among the early members of the St. Louis Gonzaga Church in Utica, which most Lebanese attended. Gabriel's son John recalled the village as "a tranquil, friendly and charming place" where he "truly grew up Lebanese as if in a transplanted Lebanese village from Mt. Lebanon." He grew up to be one of the leading scholars of the Syro-Lebanese experience in America. (Eugene P. Nassar.)

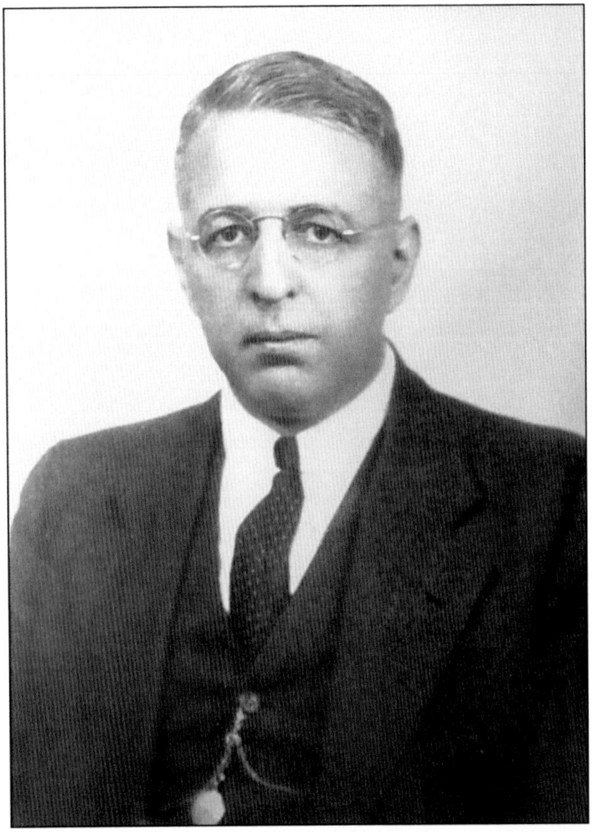

Herbert M. Hart was elected mayor on the Citizens Party ticket in 1929. His progressive leadership, which embraced all portions of the community, brought much-needed improvements in infrastructure and community solidarity even during the difficult years of the Depression. Among his lasting contributions were a park, paved sidewalks, and other improvements. (OCHS.)

One of Mayor Hart's major achievements was the creation of Pulaski Park as a public area for gatherings and recreation. There was considerable opposition to the move, but Hart managed to gather enough support among members of the village board to approve the measure. This plaque commemorates the individuals involved, including Hart and Joseph Piszcz, president of the Independent Citizen's Club who supported Hart's election. (Authors' collection.)

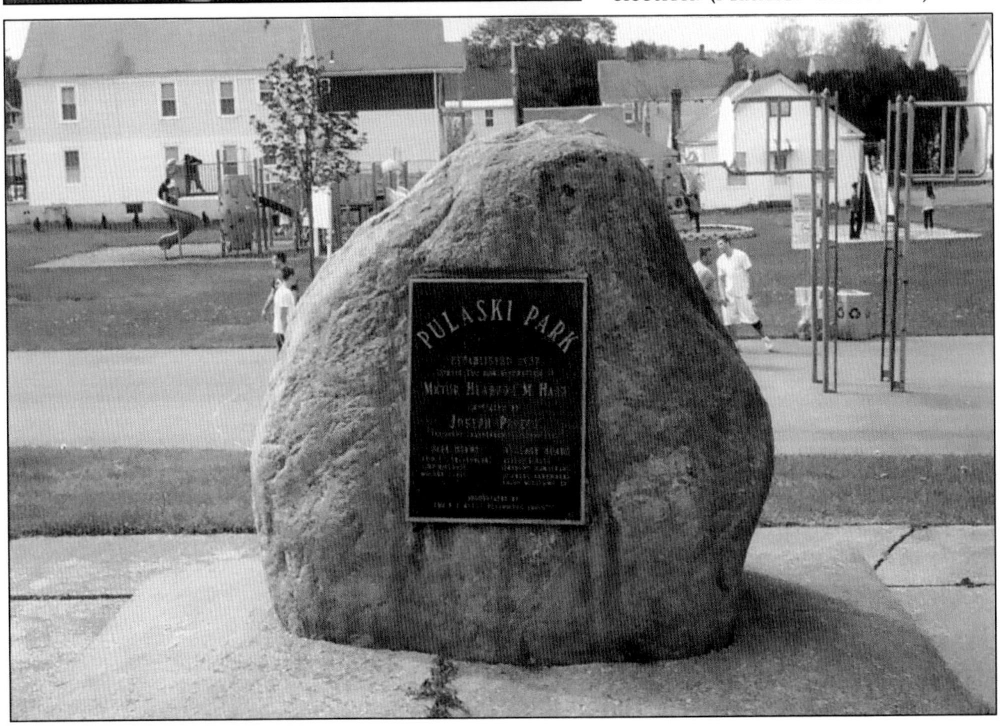

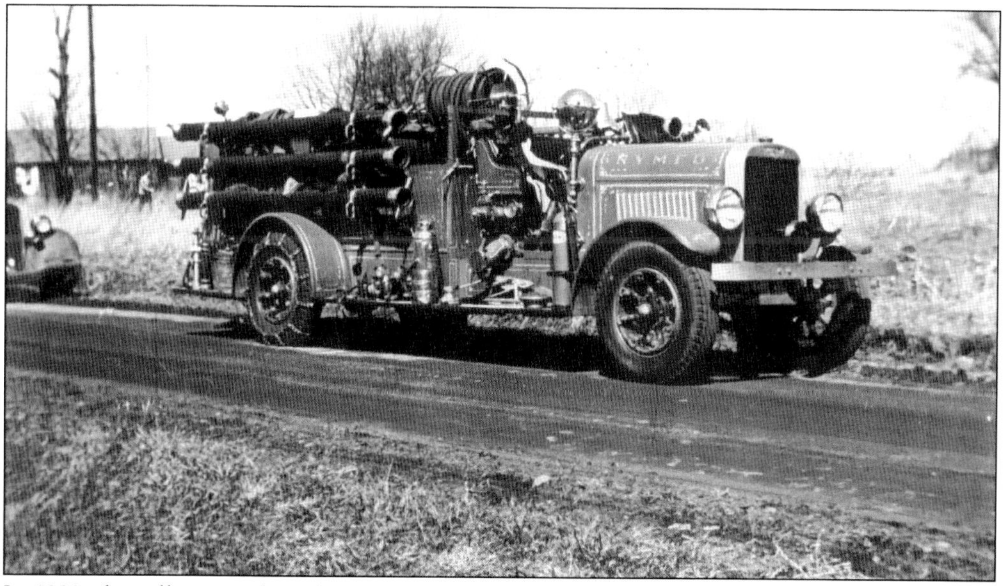

In 1929, the village purchased its first fire apparatus, a 1928 Sanford Pumper. Since there was no fire station, it was kept in the nearby Yorkville firehouse. In the following year, the village board voted to purchase the New York Mills Corporation's Community Club for $15,000 and to construct an addition to its front to accommodate two fire engines. (NYMHS.)

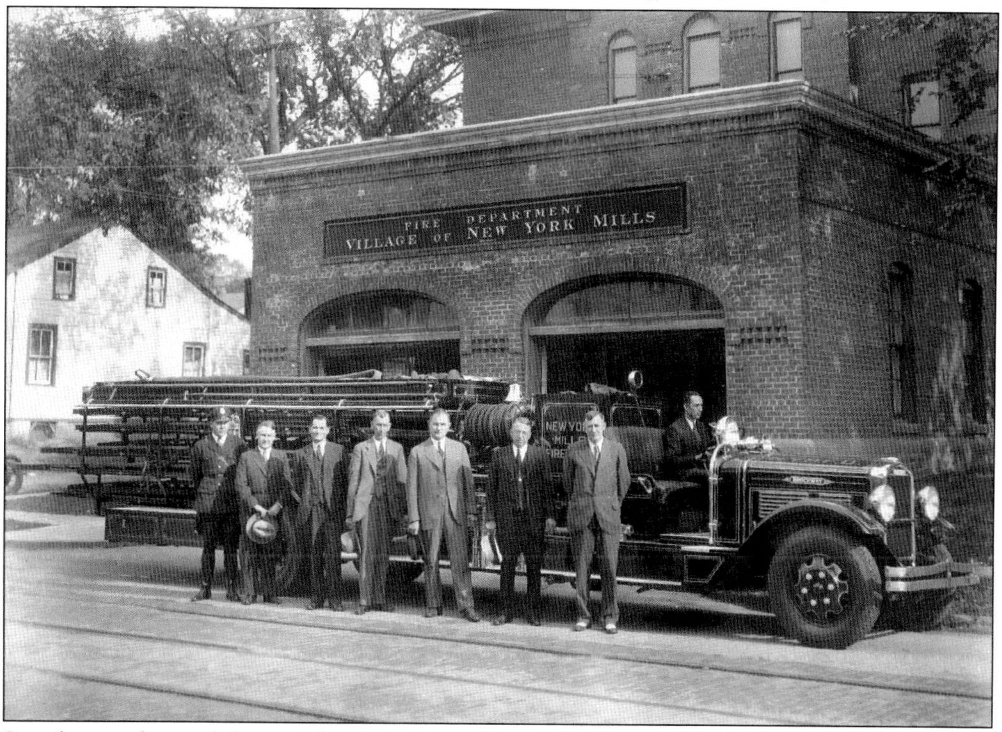

Standing in front of the new fire station with the latest acquisition of a ladder truck are, from left to right, John Kielbasa, police chief; Griff Williams, trustee; Stan Federoski, trustee; Albert Eddy, trustee; Anthony Zamiarski, trustee; Herbert Hart, mayor; Frank Bean, fire chief; and an unidentified person behind the wheel of the apparatus. (NYMHS.)

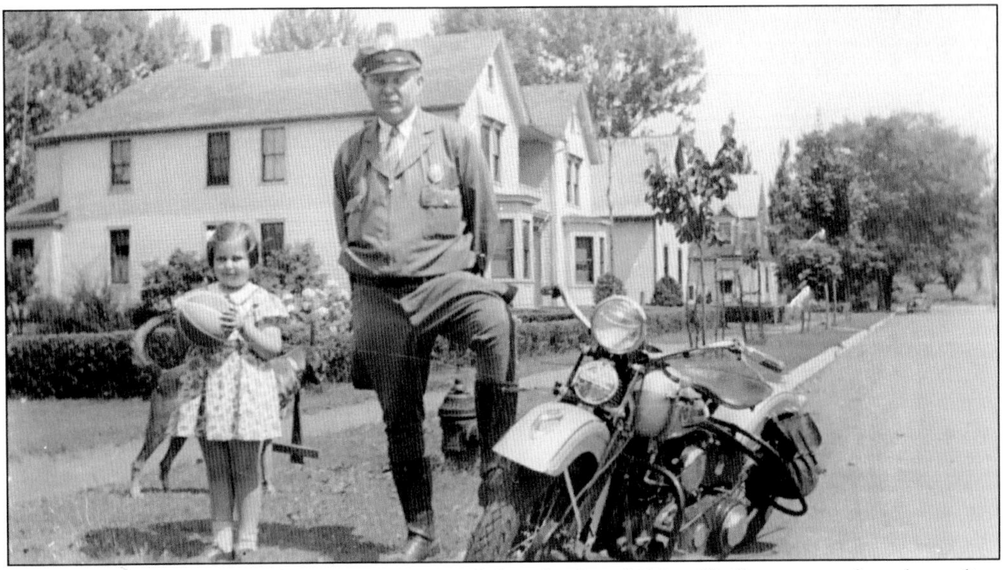

Known affectionately by the community as "John the Law," John Kielbasa served as the police chief (in fact, the only police officer in the entire village) for 25 years following his appointment by the village board in 1929. He passed away at the age of 98 in 1998. (NYMHS.)

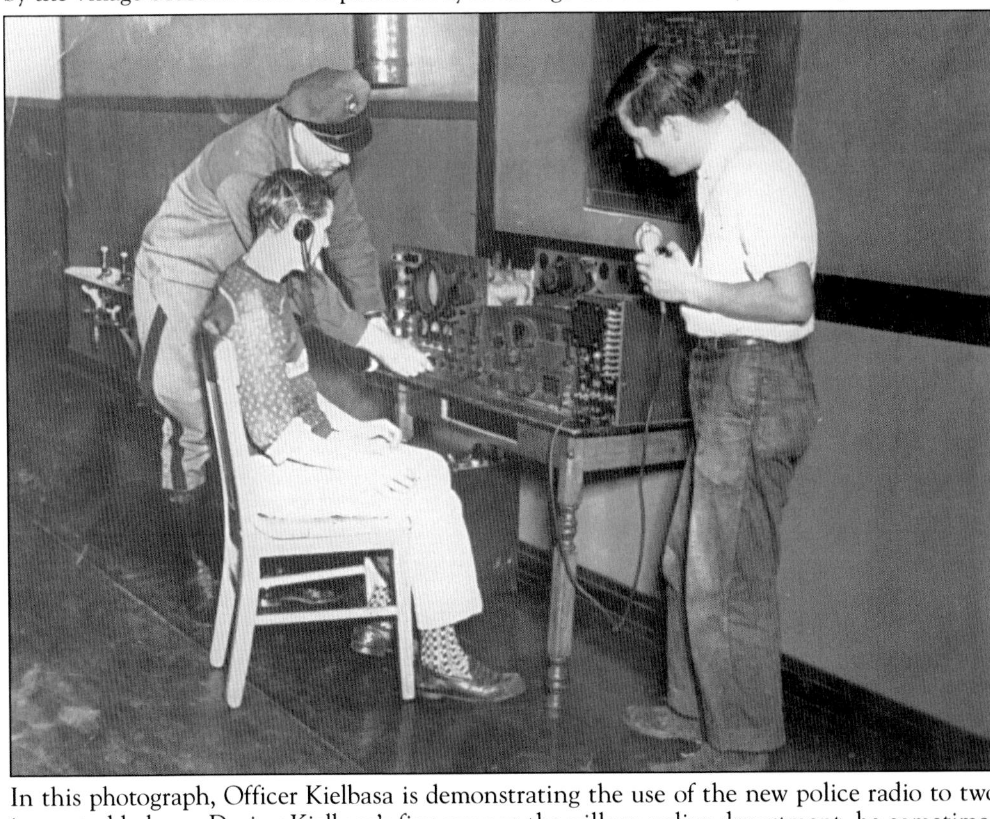

In this photograph, Officer Kielbasa is demonstrating the use of the new police radio to two interested helpers. During Kielbasa's first year as the village police department, he sometimes worked about 18 hours a day with no vacation. His annual salary was $1,300. He was not just a police officer but also a friend to everyone in the village, especially the children. (OCHS.)

This unidentified gentleman, probably one of the company officials, is seen at work inside the offices on Main Street. From here, paymasters would walk through the village streets without any guards carrying cash in pay envelopes for workers in the various facilities. Remarkably, this practice continued without incident until the company began using checks around the time of World War II. (NYMHS.)

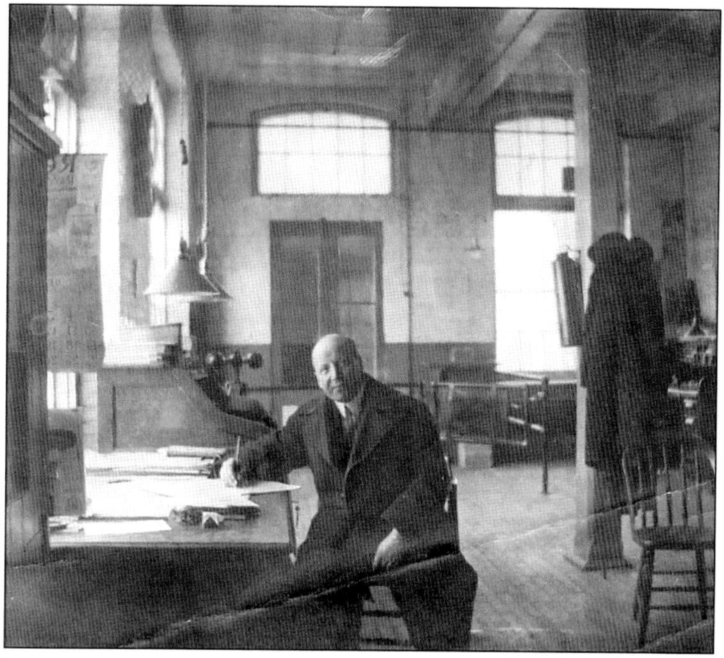

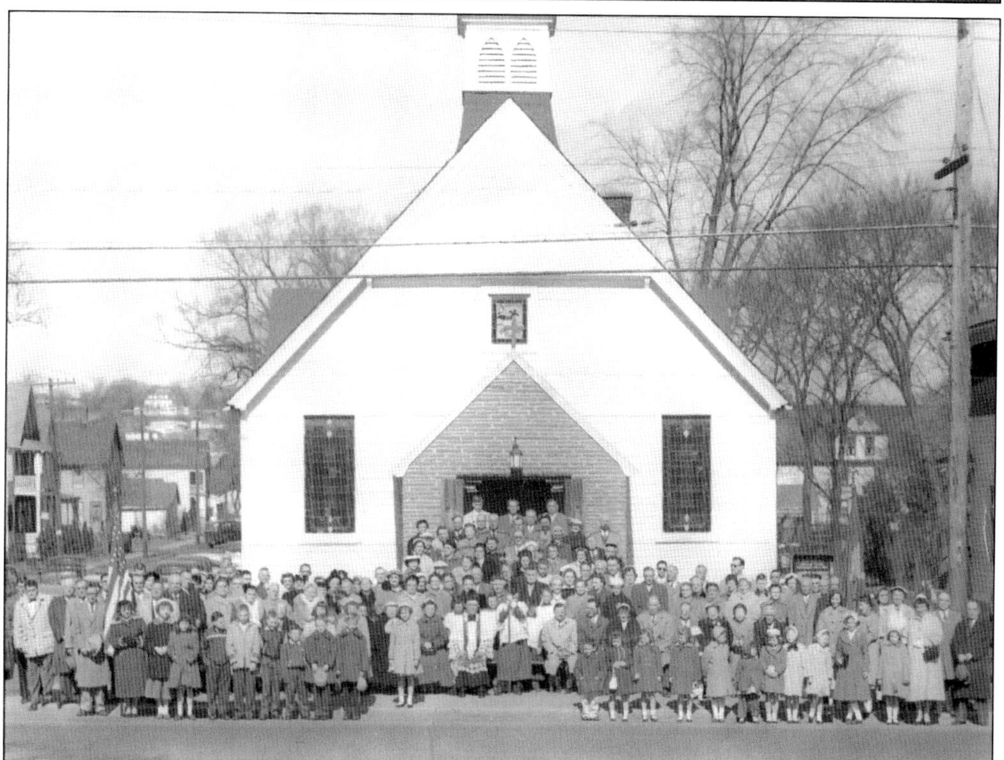

The Polish National Catholic Church was established in 1923 for Poles disaffected with St. Mary's or the Irish bishops who dominated the Catholic Church in America. The group purchased the old St. James Episcopal Church on the corner of Walcott and Main Streets along with the adjoining rectory. In this 1958 image, parishioners pose in front of their church. (NYMHS.)

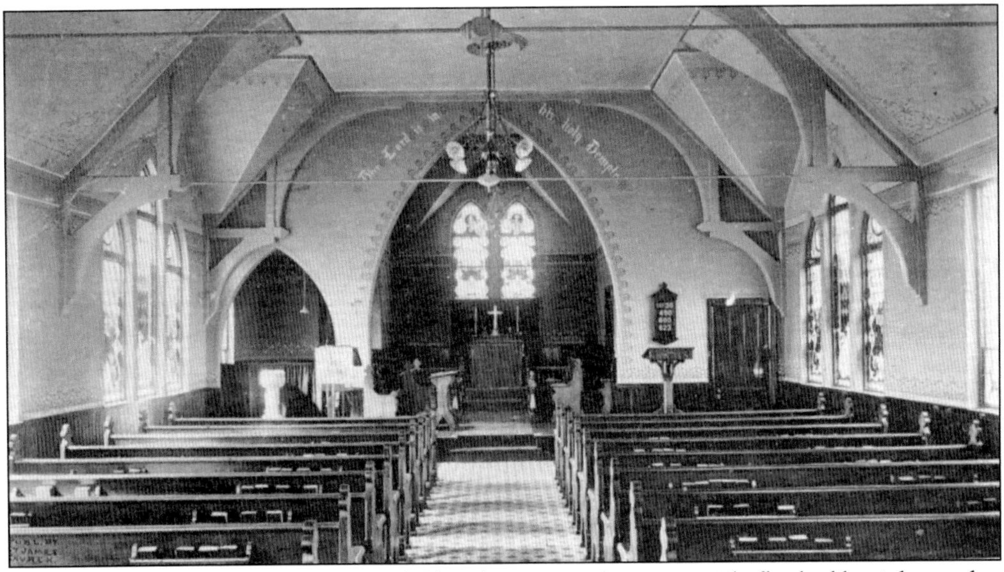

The interior of the St. James Episcopal Church contained a center aisle flanked by rich wooden pews and handcrafted windows. It was this welcoming building that became home to the Polish National Catholic Church. Much like the other places of worship in the village, the PNCC sponsored youth groups, singing societies, and community events. (NYMHS.)

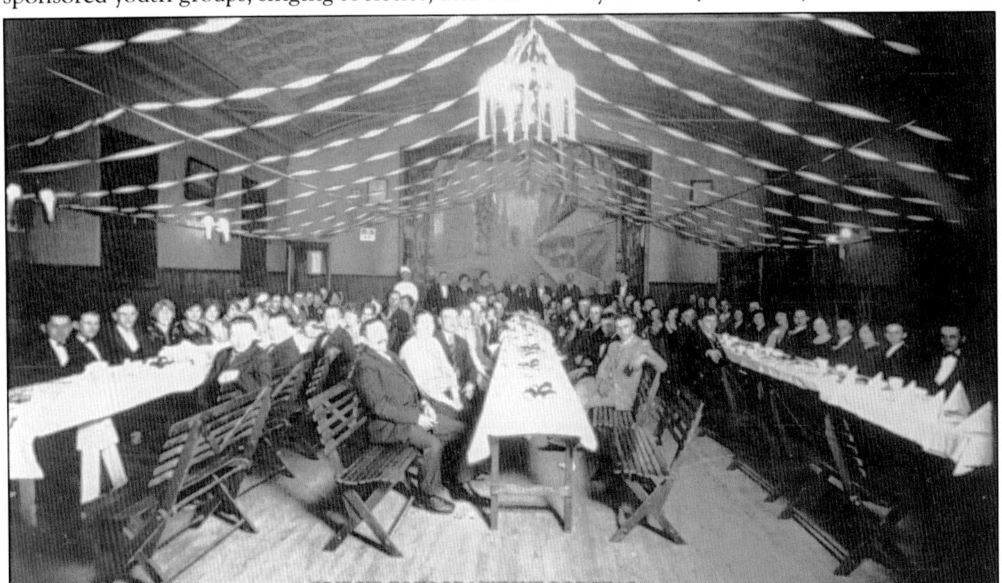

This dinner was held on November 25, 1926, on the 13th anniversary of the founding of the Polish Young Men's Patriotic Club. The organization later became a potent political influence in the village by promoting citizenship among immigrants and organizing them to vote for candidates who were friendly to the Polish population. The photograph shows the inside of Union Hall where the dinner was held. (Miller.)

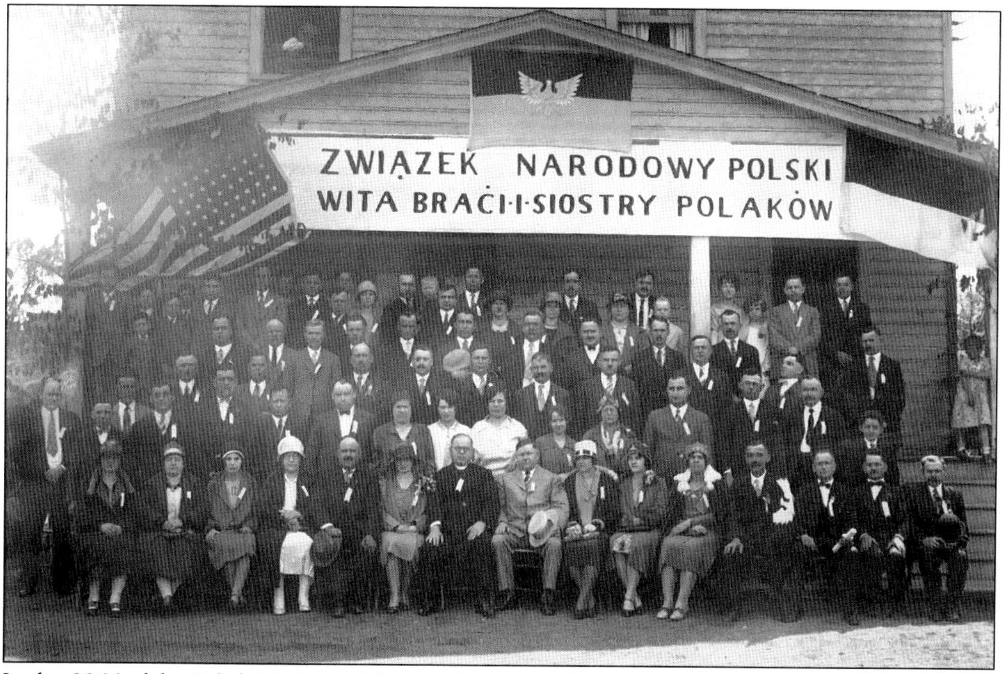

Lodge 2066 of the Polish National Alliance, known as the Progressive Patriotic Society, organized on September 4, 1918, as a chapter of a national fraternal organization offering low-cost insurance and other benefits. The banner says in Polish that the Polish National Alliance welcomes its Polish brothers and sisters. Its first president was Michael Tuman. (Authors' collection.)

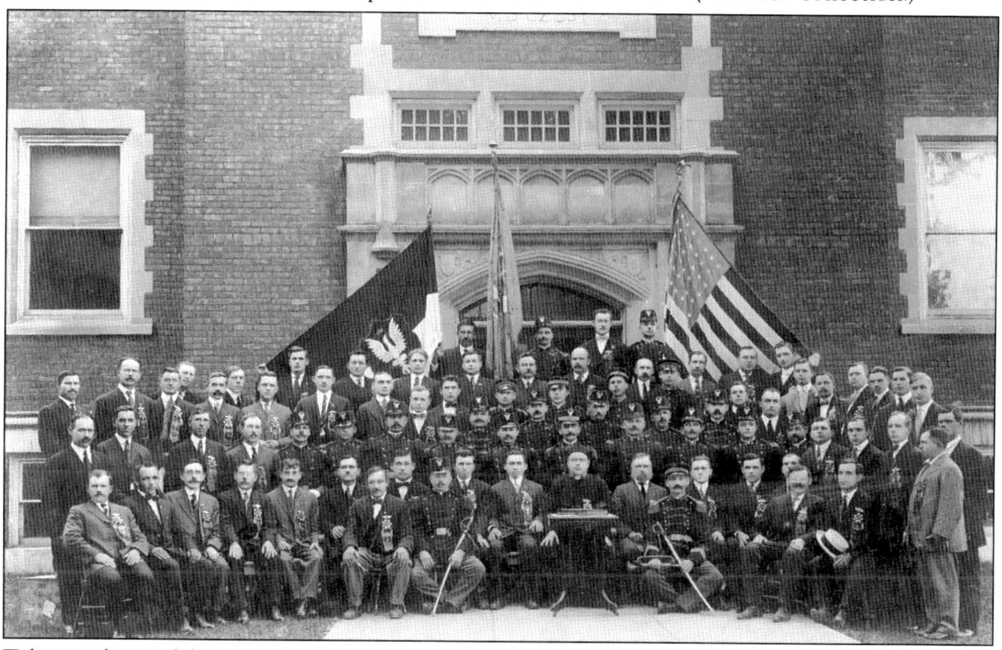

Taken in front of the St. Mary's parish school with Reverend Fijalkowski seated in the center, this photograph is believed to show the St. Casimir's Polish Men's Society. Organized in 1911 under its first chairman, Joseph Wdowiak, the group's goals were to promote brotherly love, mutual assistance, and religious projects. (Authors' collection.)

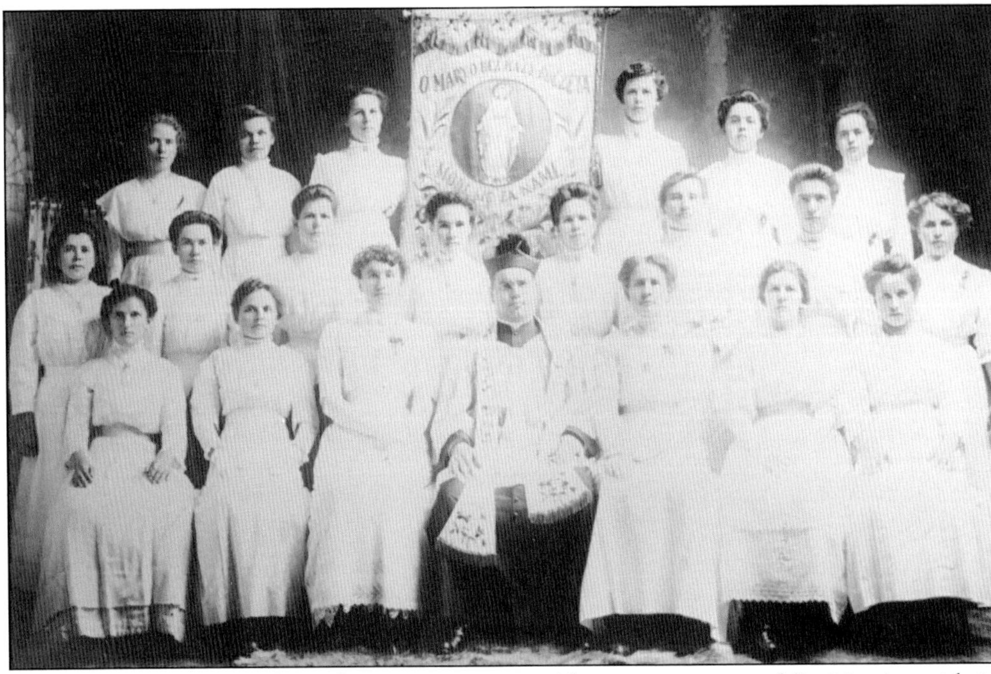

The Society of the Children of Mary was organized by young women of St. Mary's parish in December 1911. Its purposes were to provide spiritual guidance to members, participate in parish activities, and raise funds for the purchase of religious statues and other items to outfit the new church. (Authors' collection.)

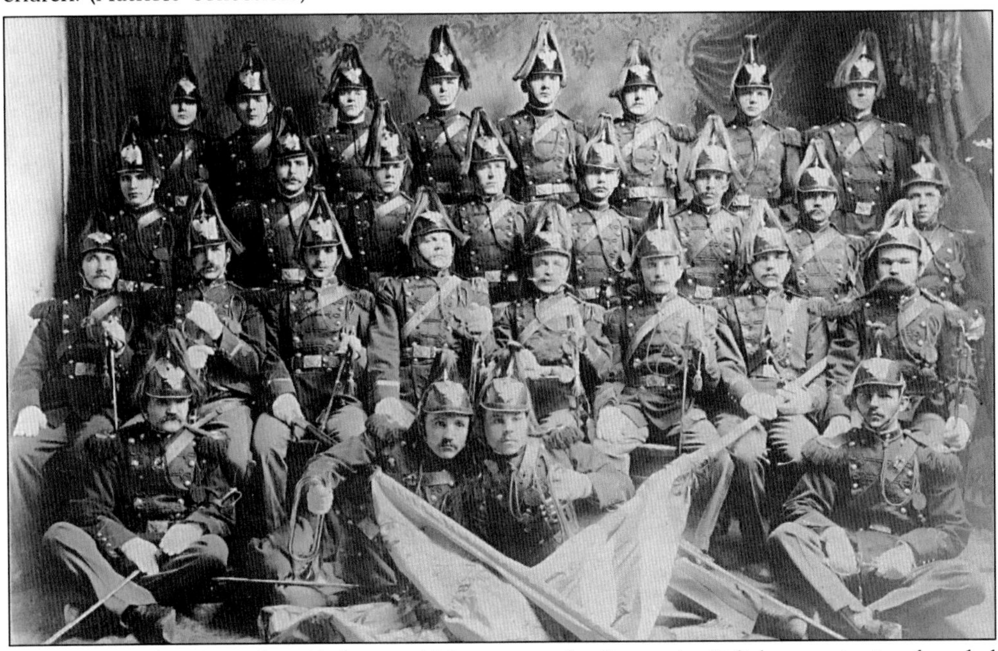

The Society of St. Stanislaus, Bishop and Martyr, was the first major Polish organization founded in New York Mills. Organized on June 25, 1905, under the leadership of Joseph Smola, its initial purposes were to collect funds for the construction of a church and to provide mutual assistance and financial help to members in times of sickness or death. (Authors' collection.)

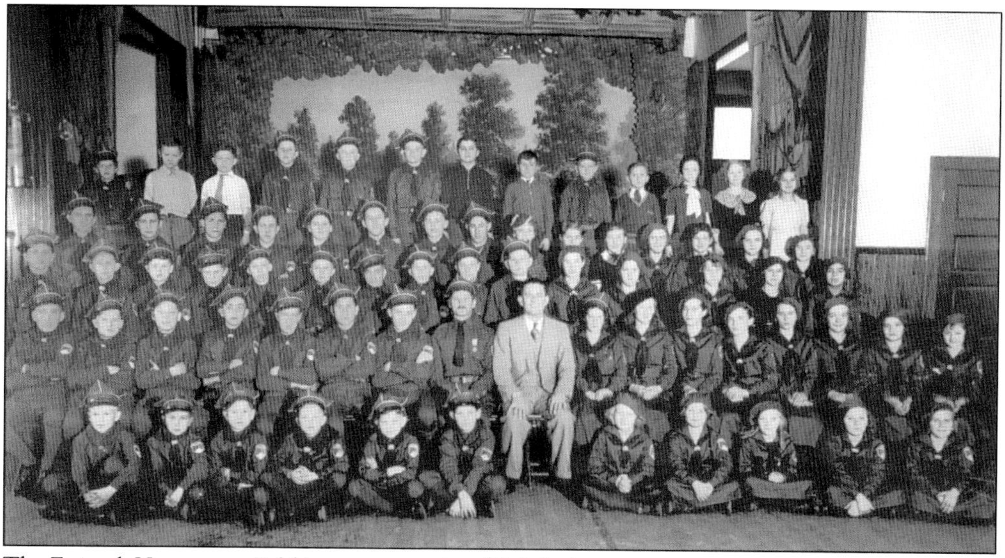

The Zwiazek Harcerstwa Polskiego was a national Polish American organization to promote scouting among young people. New York Mills had a very active program for boys and girls as this photograph illustrates. During the 1930s, Michael Dziedzic became a district leader in the movement. Note the traditional four-cornered Polish design for the boys' hats. (Authors' collection.)

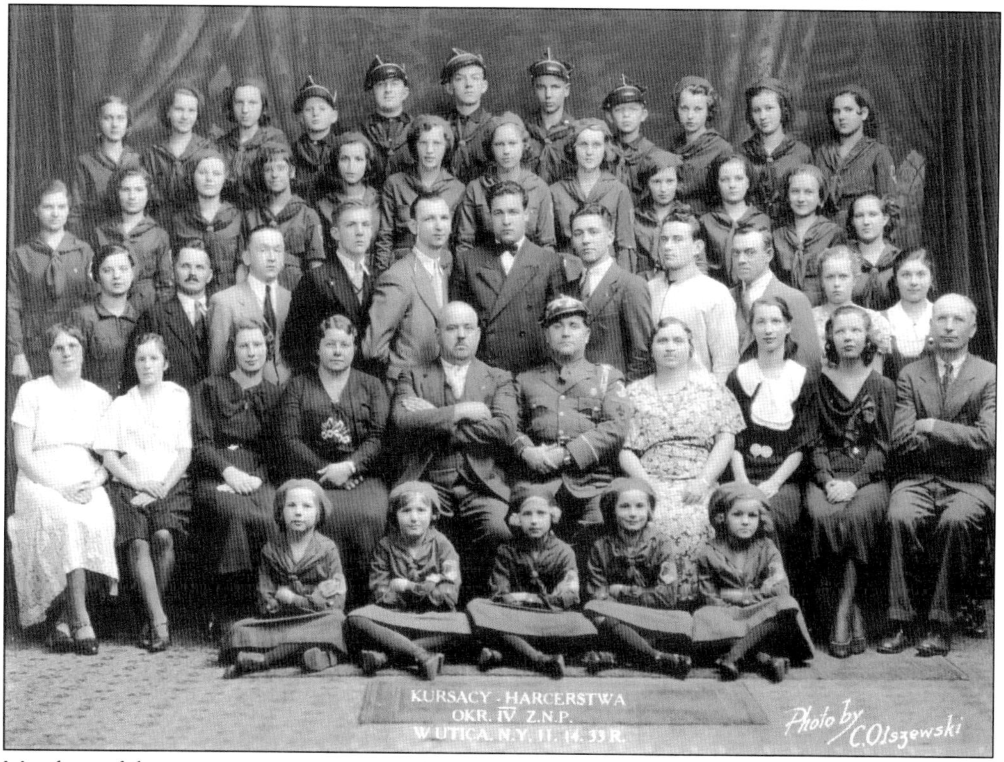

Members of the scouting troops sponsored by Lodge 2066 of the Polish National Alliance attended the regional gathering of scouts in nearby Utica and posed for this photograph on November 14, 1933. Scouting was open to both boys and girls with priority given to physical fitness and cultural and educational activities. (Authors' collection.)

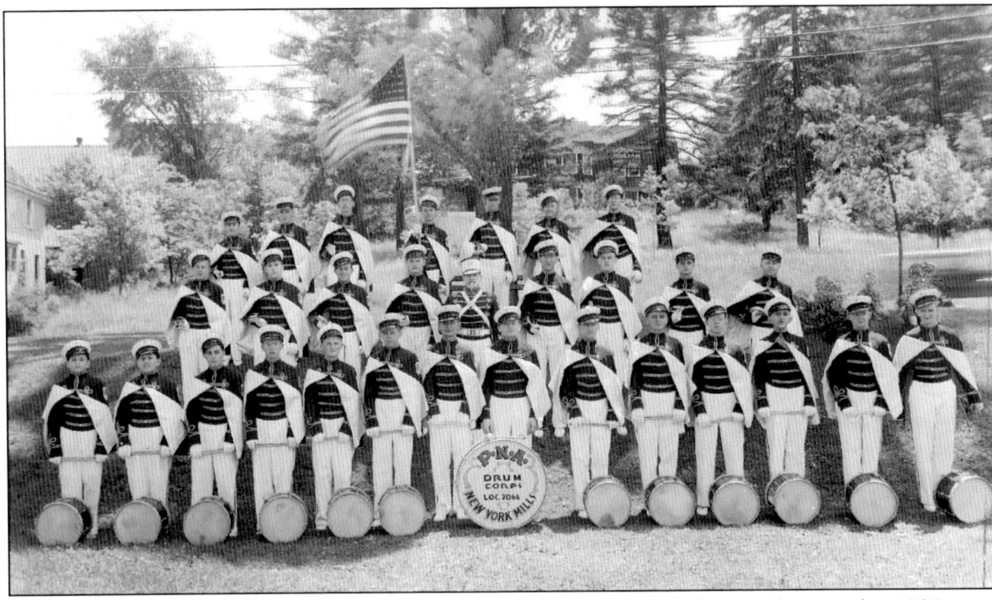

One of the other major activities other the Polish National Alliance during the 1930s was sponsorship of this drum and bugle corps that represented Lodge 2066 in parades and other events throughout Central New York. Its drum major was Edward Kwiecinski. The unit finally disbanded when so many of its members were called to military service during World War II. (Authors' collection.)

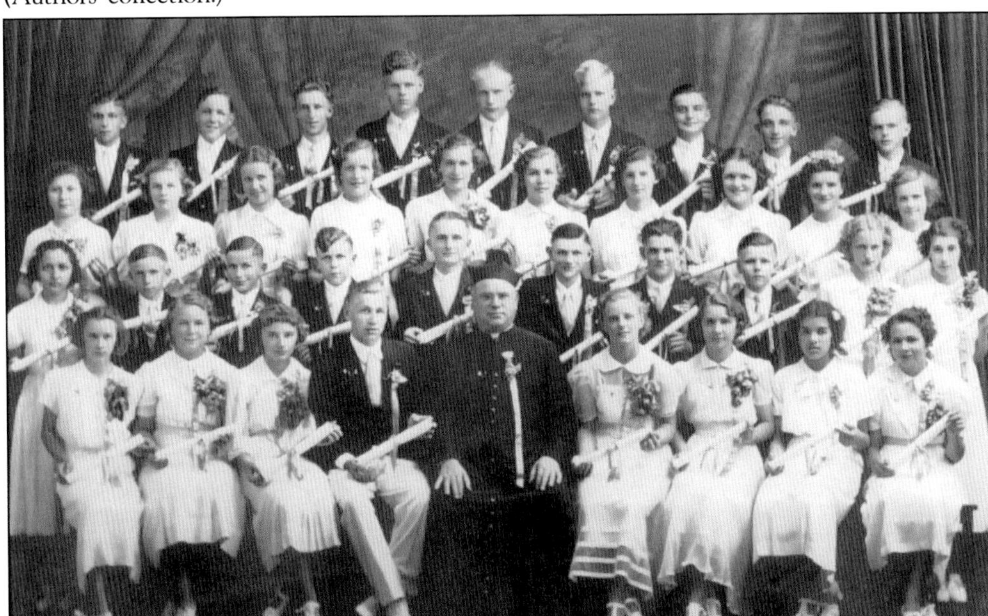

Eighth grade graduates of St. Mary's Elementary School posed for this photograph with their pastor in the mid-1930s. More than 90 percent of the village's Polish children attended St. Mary's, which enrolled more students in the 1920s and 1930s than did the public elementary school. In 1930, it had an enrollment of 494 students taught by nine Felician nuns, a ratio of 55 to 1. (Authors' collection.)

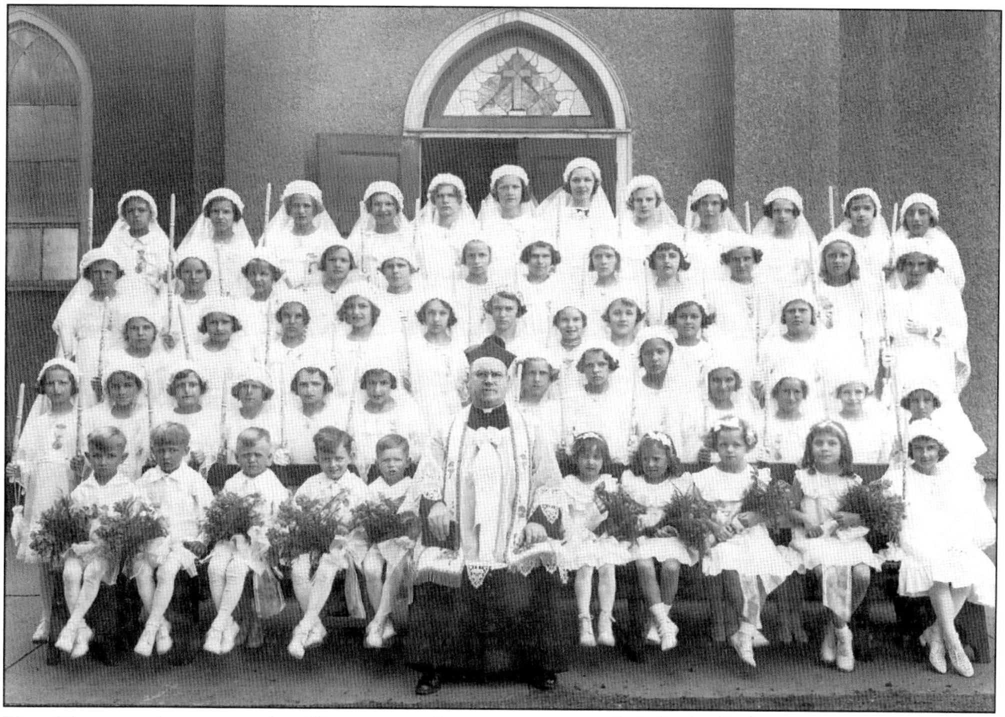

First Communion was a significant rite of passage among Polish Catholics as evidenced by the formal attire worn in these two photographs from St. Mary's parish in the 1930s. Following the church ceremony, it was traditional for the parents to invite family and friends for a meal to celebrate the occasion. (Both, authors' collection.)

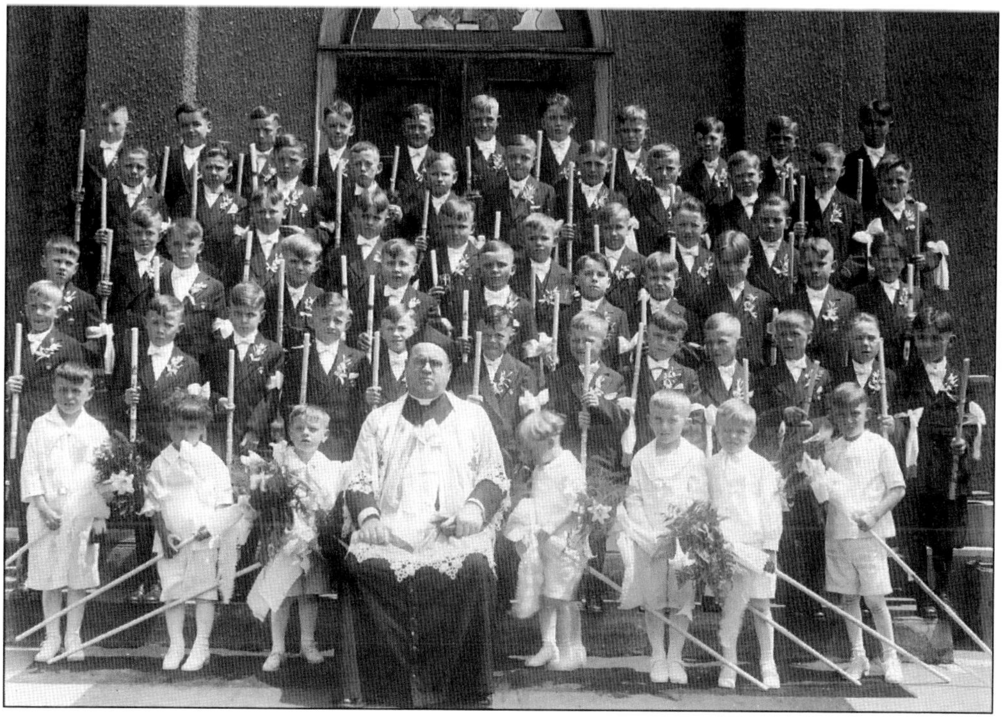

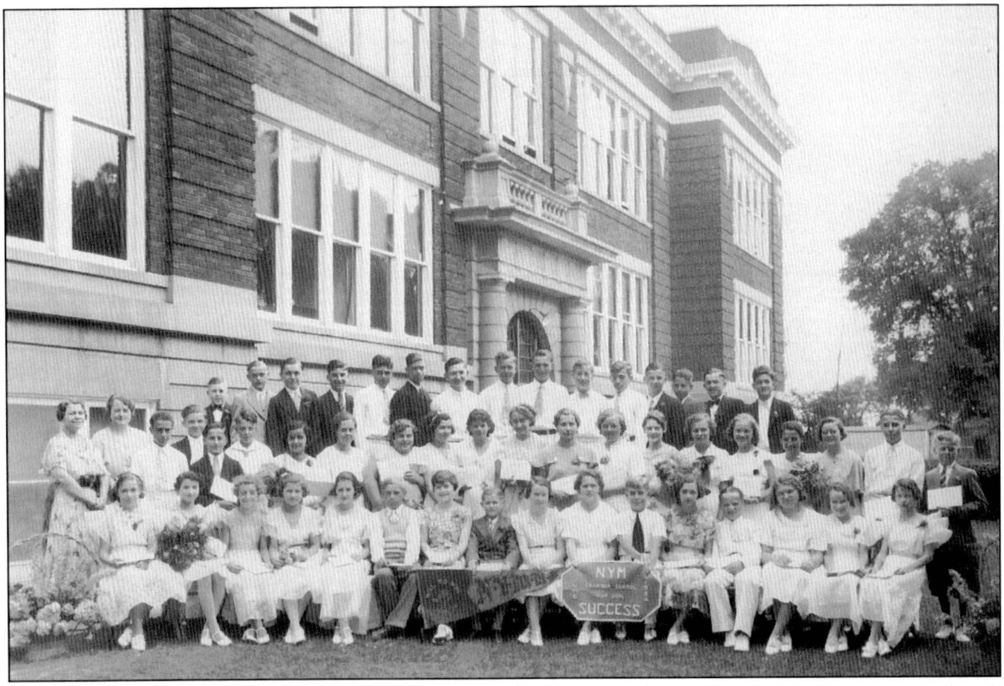

"Success" was the motto of the graduating class of 1934 at the New York Mills Grammar School. The school pennant carries the slogan "Will to Win." At the beginning of the following school year, 406 students enrolled in the elementary school and 224 in high school, all in what was even then considered an overcrowded building. (Authors' collection.)

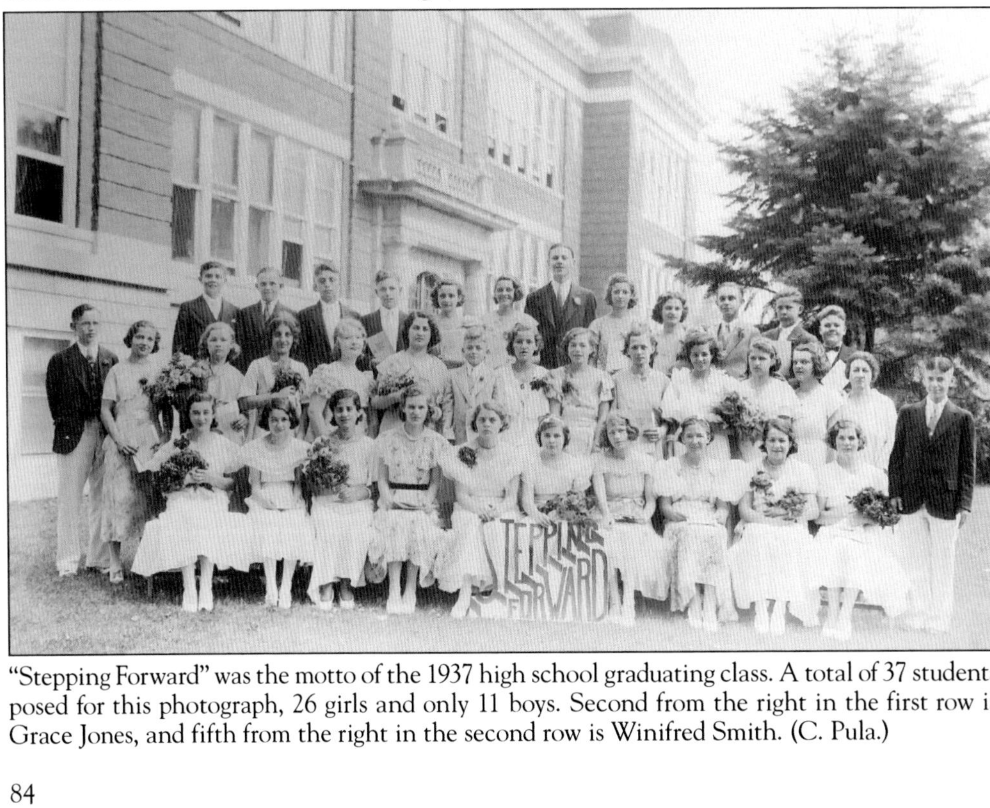

"Stepping Forward" was the motto of the 1937 high school graduating class. A total of 37 students posed for this photograph, 26 girls and only 11 boys. Second from the right in the first row is Grace Jones, and fifth from the right in the second row is Winifred Smith. (C. Pula.)

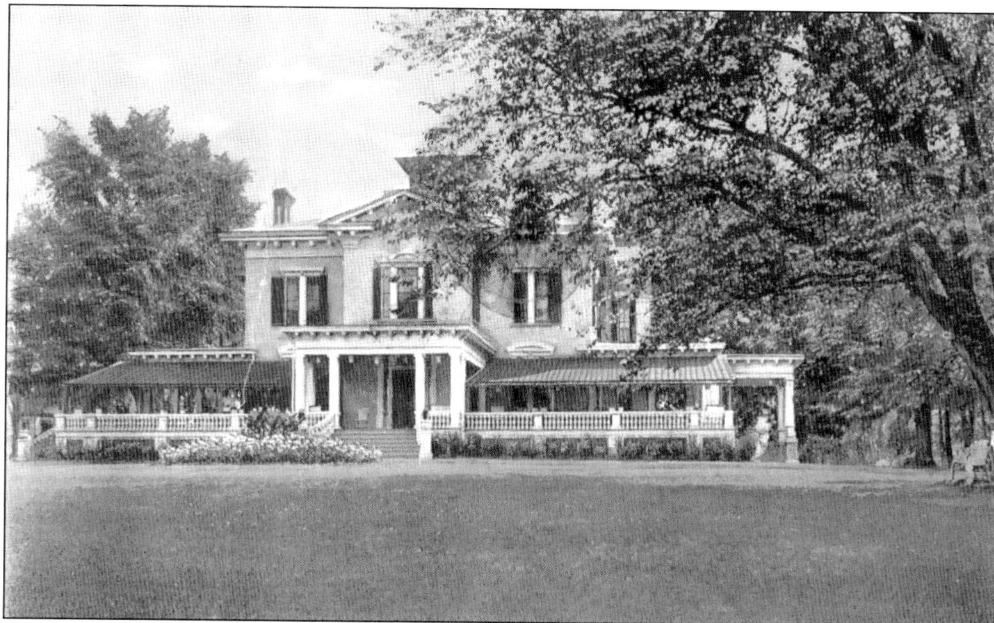

The Utica Golf Club, located on the former Samuel Campbell estate, was purchased by Arthur Hines and Patrick Casey who renamed it the ArHiPaCa Golf Course using the first two letters of each name. The former Campbell home became the Castle Inn at the course for many years. The course gave local boys an opportunity to earn money caddying. (OCHS.)

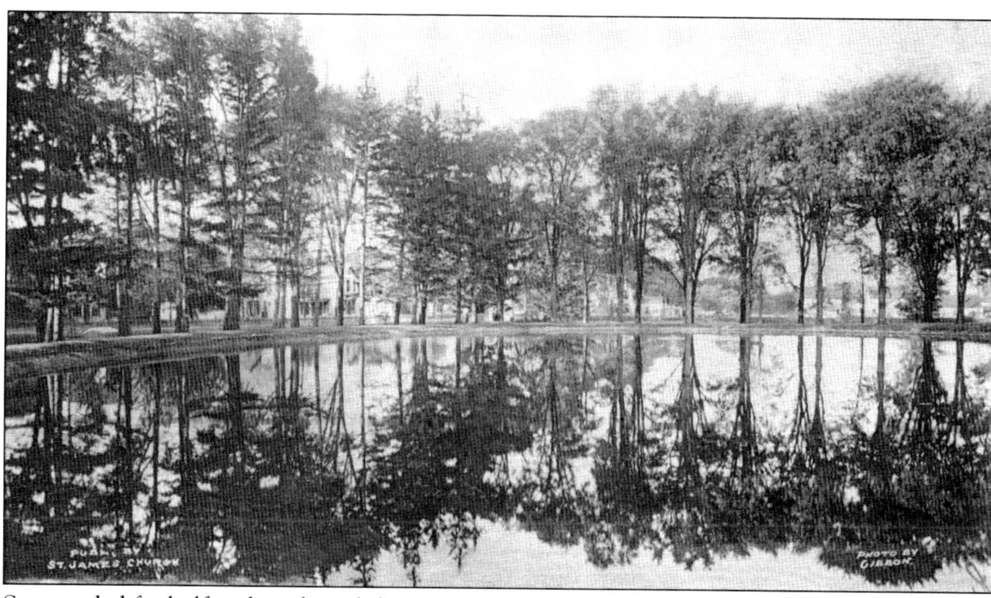

Surrounded for half its length with beautiful shade trees, on a clear day the lower pond on the Campbell estate often produced the mirror image seen in this photograph. Although attractive to behold, when the pond was incorporated into the golf course it became instead a nightmare for many golfers. (Pytko.)

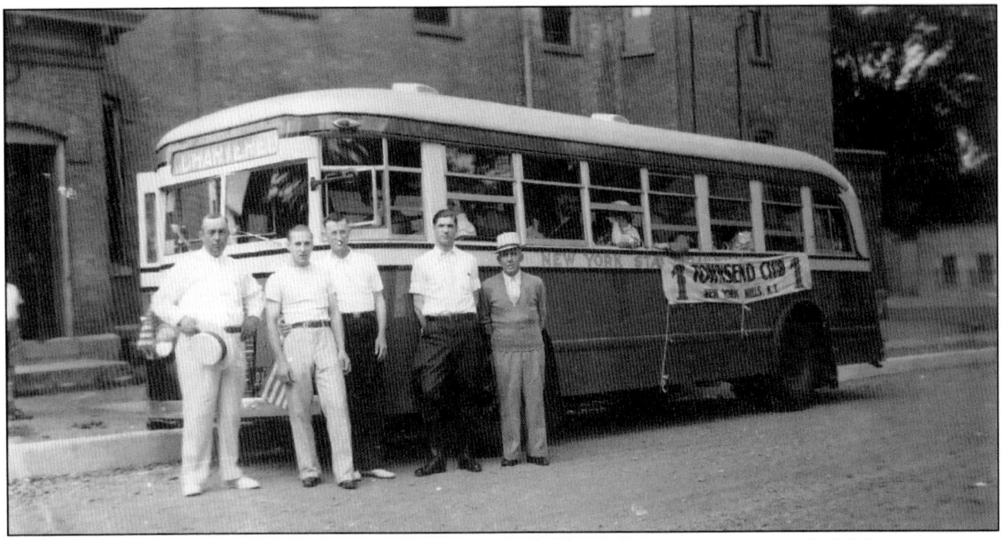

The Townsend Club, named after a politician who lobbied for senior citizens, held dinners, bake sales, and raffles to raise money for various causes. The club also sponsored outings to nearby parks and other places of interest. Pictured here are Amos Smith (left), Fred Healey (second to left), and other leaders of the organization. (C. Pula.)

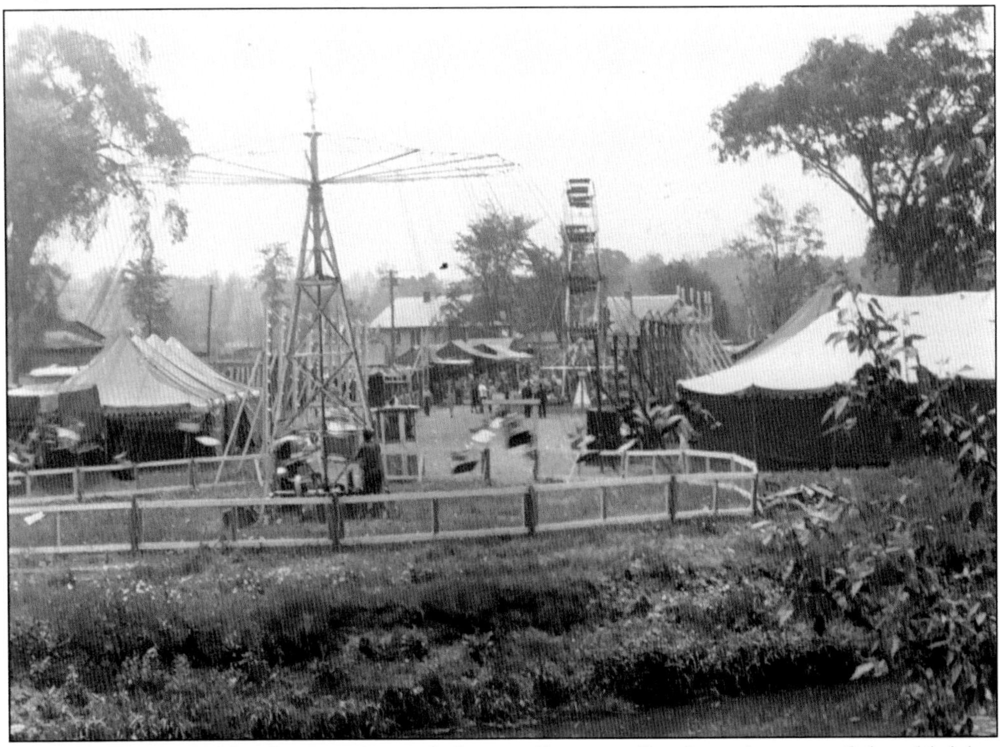

Traveling carnivals like this one, located close to Sauquoit Creek in the area of the old dykeside park, provided summer entertainment for young and old alike. Clearly visible here are a Ferris wheel, other rides, and tents for games and refreshment. Traveling musicians and other performers came and went through the area adding some variety to leisure activities. (Kozak and Unczur families.)

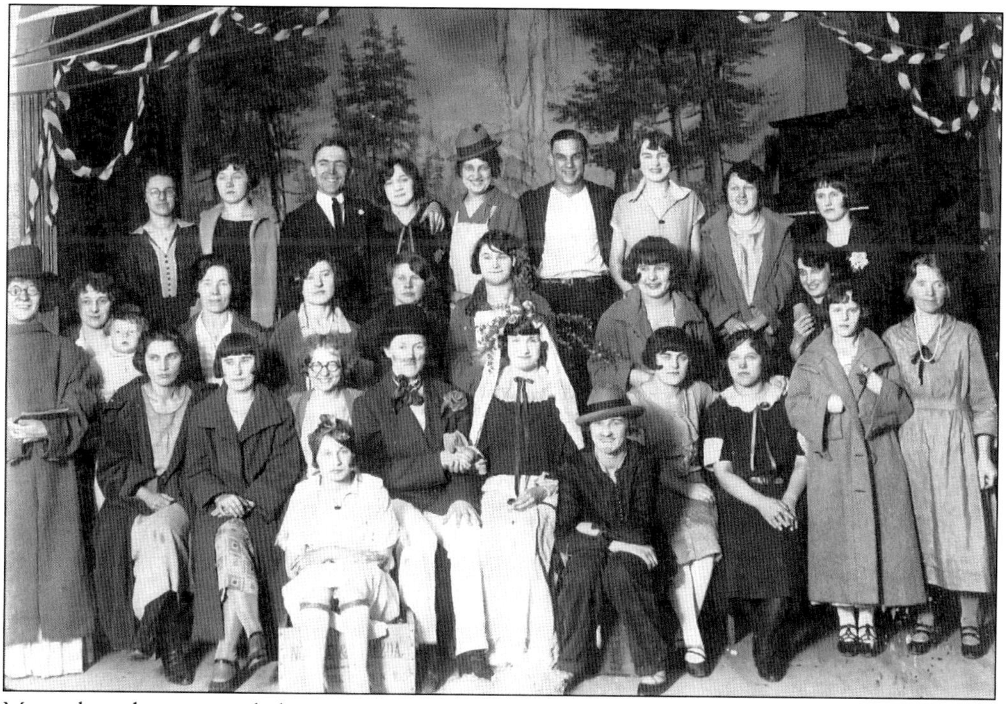

Musical productions and plays were popular pastimes in the years between the world wars. This group posed for their photograph in costume for a production in Union Hall, which was a favorite venue with a relatively large stage and open floor space that could be arranged in any configuration that might be needed. Popular themes included classical productions and those that dealt with Polish historical topics. (Authors' collection.)

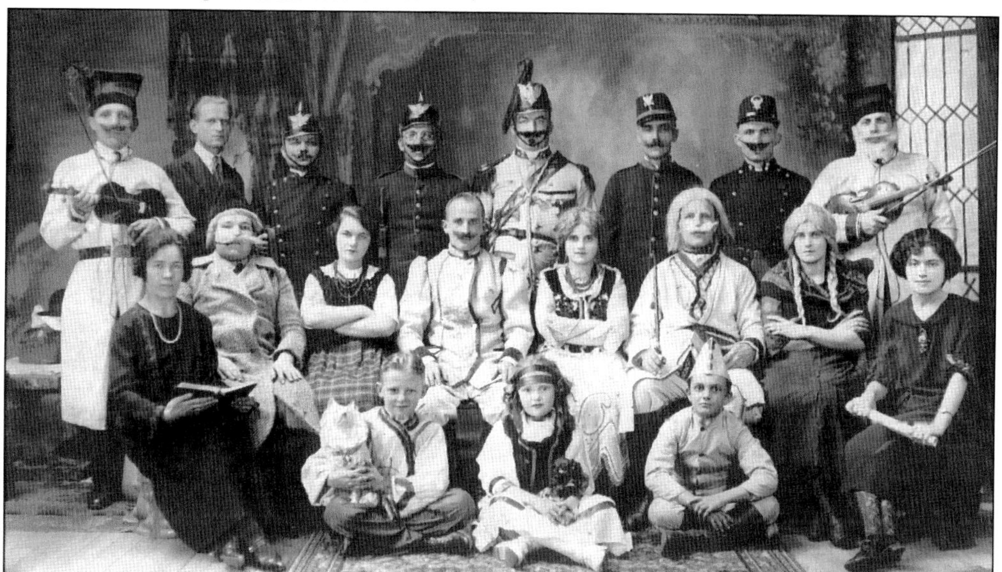

The first parish play on April 23, 1922, included in the second row Mr. Sadlik (second from left), Joseph Romanow (fourth from left), and Barbara Tuman Jakubowski (far right) and in the third row Jan Pula (second from left) and Joseph Klimaj (third from right). In addition to religious performances, the parish also sponsored plays on Polish patriotic and cultural themes. (Berdell.)

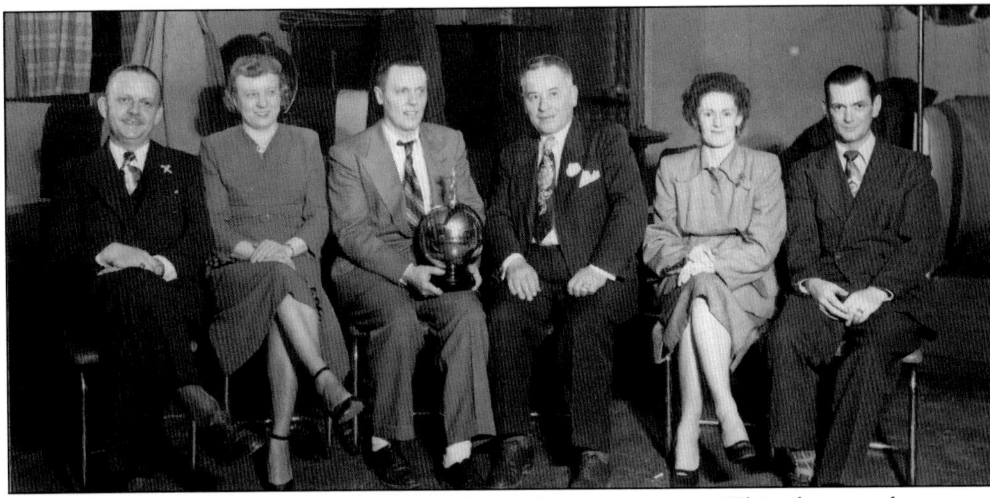

Bowling was a popular pastime among adults in the postwar years. This photograph captures the members of the Marine Bar team, champions of the Gutter Ball Coed Bowling League in 1949–1950. Marine Bar was located on Burrstone Road in a building that later became the Chowder House restaurant. Holding the trophy is Frank Stanek. (NYMHS.)

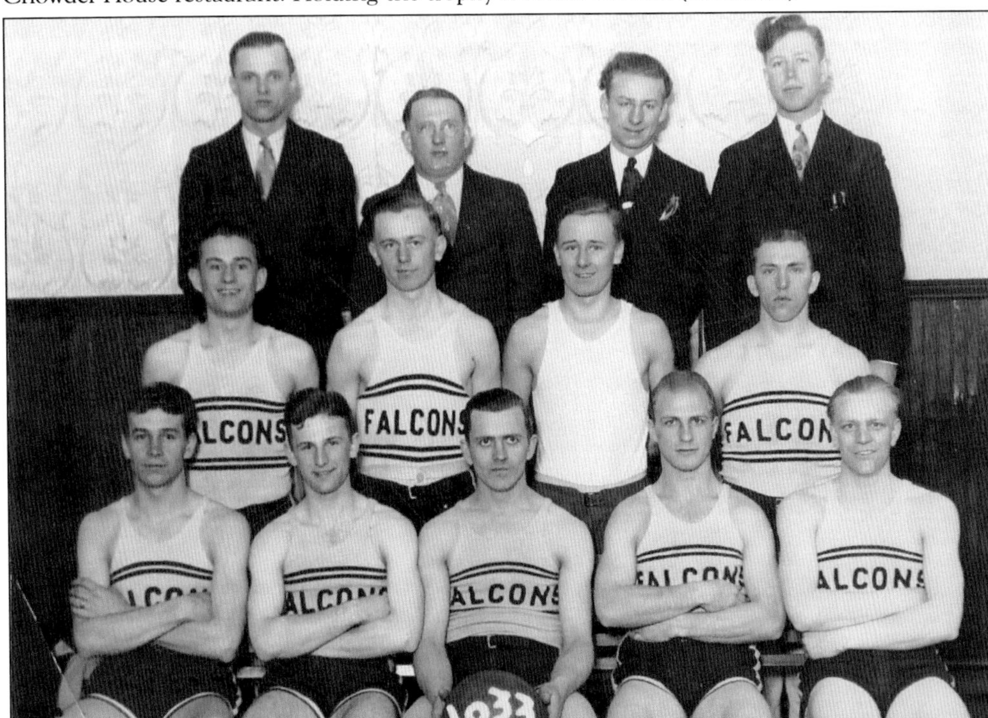

The Polish Falcons Alliance was a national organization somewhat similar to the German Turnverein or the American YMCA in stressing physical fitness. Nest 320 was founded in New York Mills in 1912 with Piotr Kozak as its first president. The Falcons sponsored various athletic teams including this 1933 basketball squad. (NYMHS.)

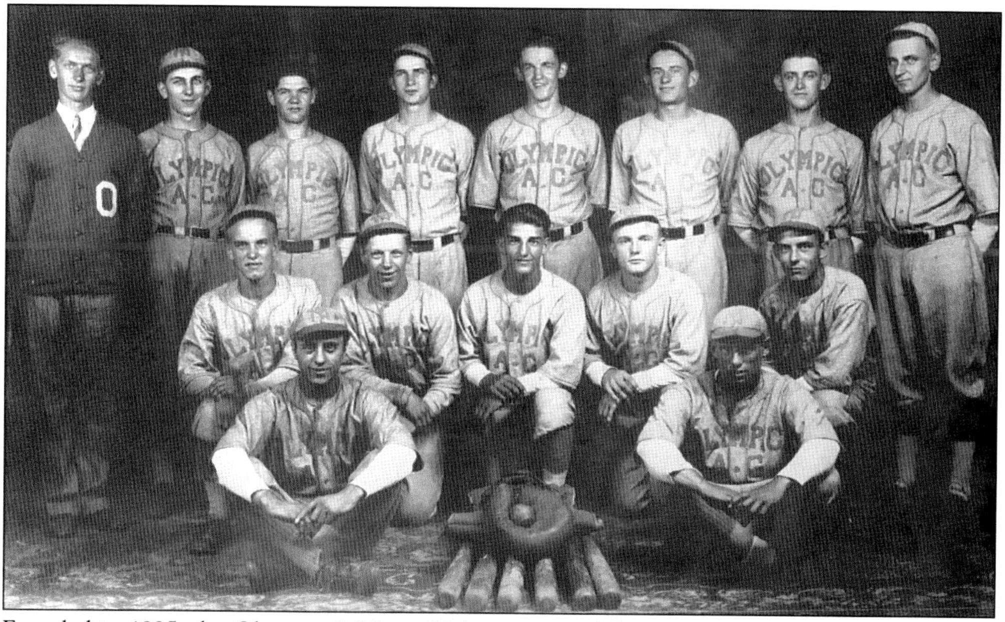

Founded in 1925, the Olympic Athletic Club was one of the most successful baseball teams in central New York. From the left to right are (first row) Christy Zampardi and Matt Furgal; (second row) Zigmund Chrabaszcz, Stanley Lachut, Matt Soja, Walter Burak, and Andy Wolak; (third row) John Trzepacz, Joe Kogut, Matt Lachut, Ben Swierat, Mike Burak, Mike Dziedzic, Walter Nogas, and Stanley Ryczek. (Authors' collection.)

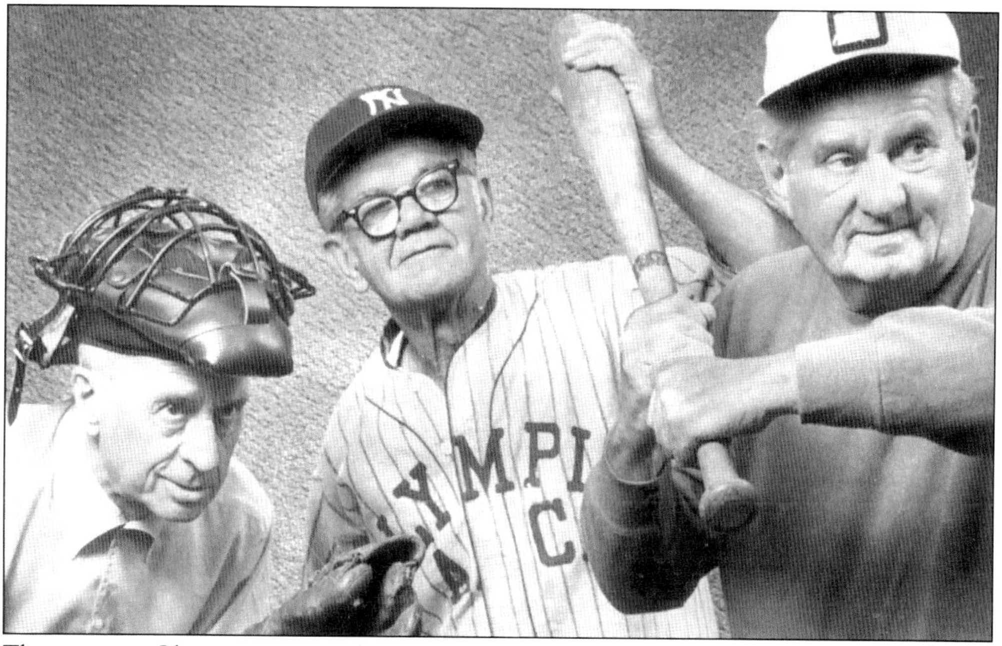

The surviving Olympic team members held a reunion at Alexander's Inn in Whitesboro on their anniversary in 1981. Taken on that occasion, this photograph shows original team members, from left to right, Christy Zampardi, Matt "Perchy" Lachut, and Michael Dziedzic. In many respects, it was also symbolic of the bygone era when the village fielded multiple neighborhood teams in several sports. (Authors' collection.)

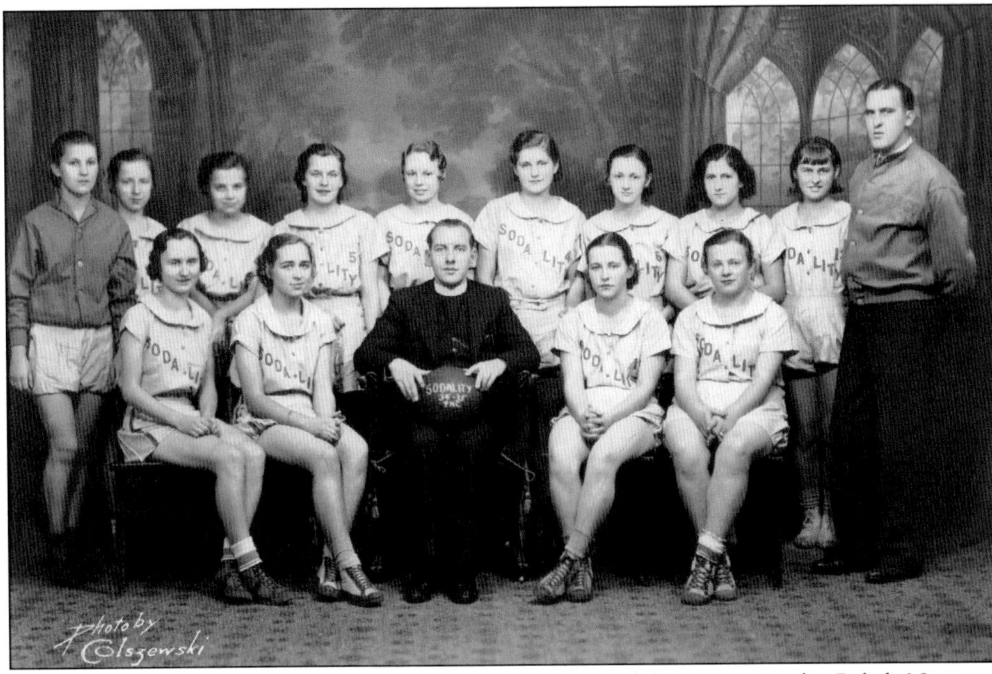

This 1934–1935 basketball team was sponsored by the Sodality group at the Polish National Catholic Church. Pictured here are, from left to right, (first row, seated) Olga Bednarz, Helen Saletnik, Rev. Joseph Czyzewski, Tillie Motyka, and Helen Jaja; (second row, standing) Jessie Rogowski, ? Tobias, ? Radlow, Frances Justyna, Emily Chrabaszcz, Frances Tymczak, Wanda Potrzeba, Josephine Niziolek, Lola Czupryna, and coach Starkey Lewis. (NYMHS.)

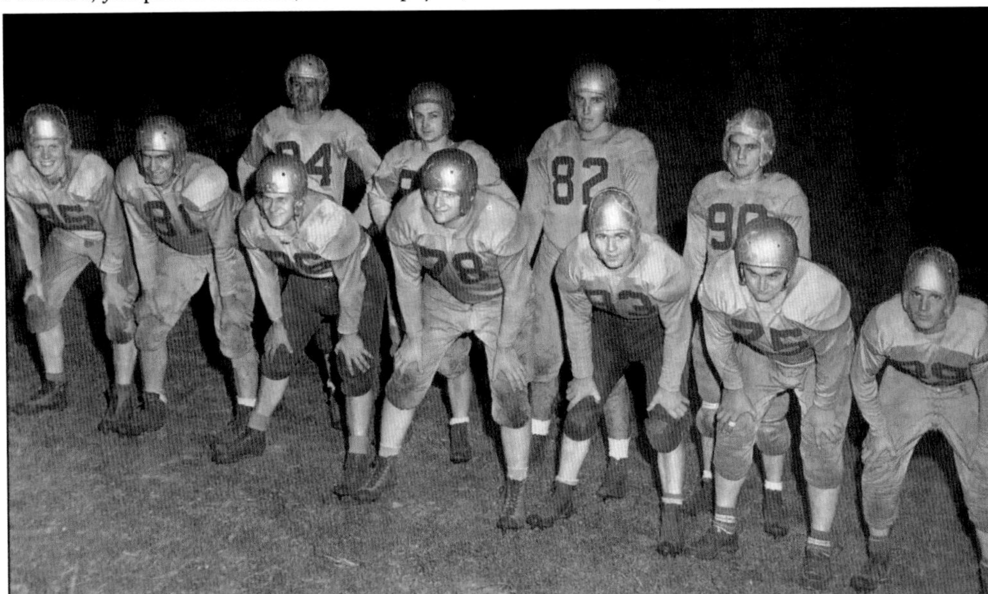

The New York Mills Terriers football team is pictured on the eve of its showdown with the Sidney Bisons. On the line are, from left to right, Don Brown, Ted Niziolek, Norm Swattski, Jack Scholl, Frank Jordan, ? Johnson, and ? Kupiec; in the backfield are Jim Chapman, Joe Torr, Dan Farrell, and Bob Farrell. (NYMHS.)

Five
War and Transformation

The Japanese attack on Pearl Harbor once more brought war to America. At home, the textile mills turned to making uniforms and other wartime necessities, people collected metal and rubber for recycling, community groups sold war bonds, and individuals planted Victory Gardens. At the high school, an unprecedented 100 percent of the students participated in the Victory Garden program, with many also raising money for war bonds. In 1943 alone, workers at A.D. Juilliard planted over 15 acres of gardens, while Group 2066 of the Polish National Alliance was commended by treasury secretary Henry Morgenthau Jr. for its purchasing of war bonds. On July 10, 1945, a Utica newspaper spotlighted New York Mills for "outstanding performances in the recent Seventh War Loan campaign." With a goal of $42,000, the village purchased $116,741.25 in bonds, 278 percent of its target.

If civilian support was outstanding, so was the contribution to the armed forces. In 1942, the *Observer-Dispatch* ran an article titled "Center of Patriotism: Single Street Sends 20 to War" about the tremendous contribution of Maple Street, a short block of four company houses and one private residence whose 10 families had already given 20 sons to the service. Two years later, the newspaper ran another article noting that some 700 village residents were already in the service, a number that grew to well over 800 by the end of hostilities.

At the conclusion of the war, the A.D. Juilliard Company rewarded returning servicemen with a $100 bonus. Another program that assisted returning veterans was the 52-20 Program. This provided $20 per week for up to 52 weeks until a vet found a job. Several village residents established a Father's Club under president Kazimierz Dziedzic to serve the village and its veterans. The group provided funds for various projects, including the World War II monument that listed the names of those who served during the war. Stanley Powroznik summed it up best: "They responded to duty in the face of the enemies and contributed more than their share to keep our freedom."

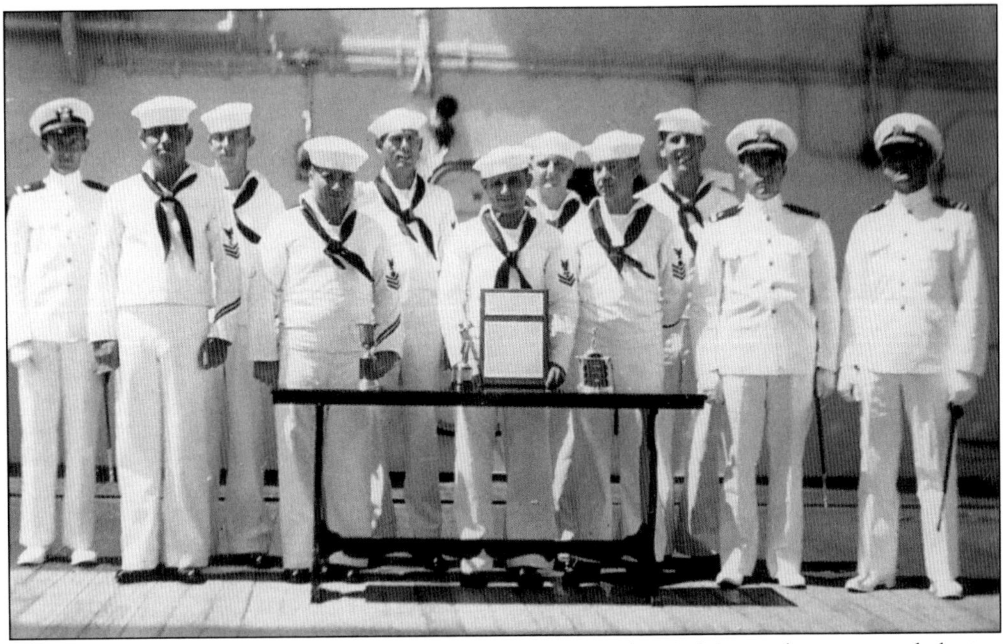

Vincent Nogas enlisted in the Navy in 1933. While stationed in California, he continued playing golf, winning the Long Beach City Championship, San Pedro City Championship, Palos Verdes Championship, and Long Beach Area Navy Championship. In 1940, he won the All-Navy Championship. He is pictured here receiving the latter award. His collection of trophies was lost in the sinking of the USS *Lexington* in 1942. (Nogas.)

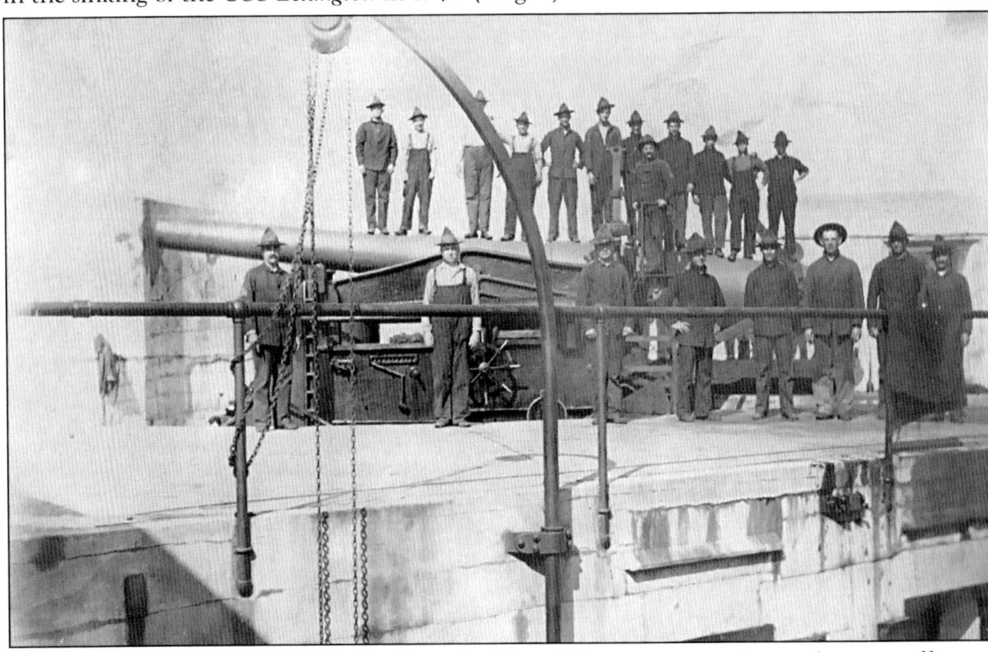

Fearing the rising tensions in the Far East, the United States attempted to make some efforts to fortify the Philippine Islands despite the isolationist sentiment in congress. In this photograph, Pvt. David Ruse (third from right) from New York Mills helps to install the first large artillery pieces on Corregidor Island in Manila Bay. (OCHS.)

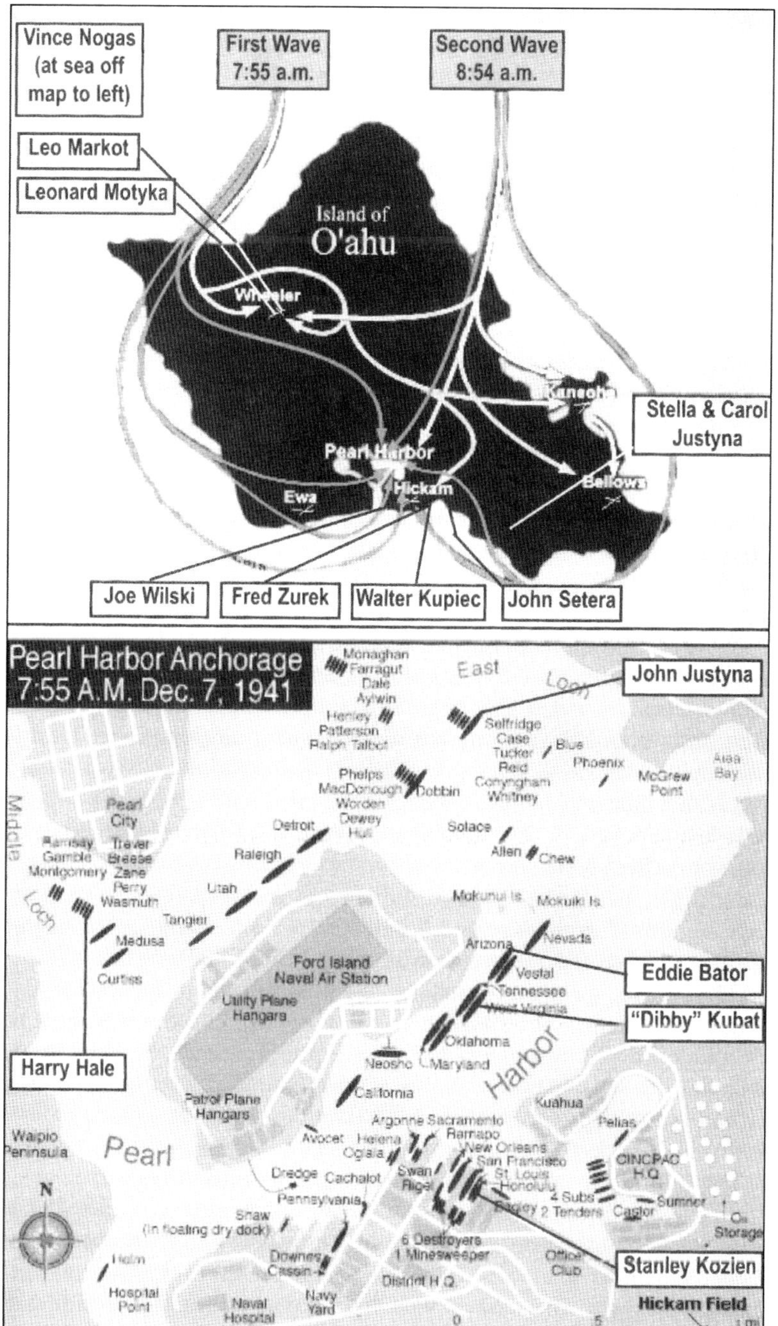

Surprisingly, considering the entire population of New York Mills was less than 3,500, there were 13 people from the village in the Hawaiian Islands when the Japanese attacked Pearl Harbor on December 7, 1941. These maps give their names and locations at the time of the attack. Five were on ships that were attacked in the harbor, four in Army installations close to the harbor, two at Wheeler Field, and two (Stella Justyna and her daughter Carol) at home in the hills overlooking Honolulu. "That day will stay with me forever," Leo Markot commented a half-century later, "I think about it almost every day." (Authors' collection.)

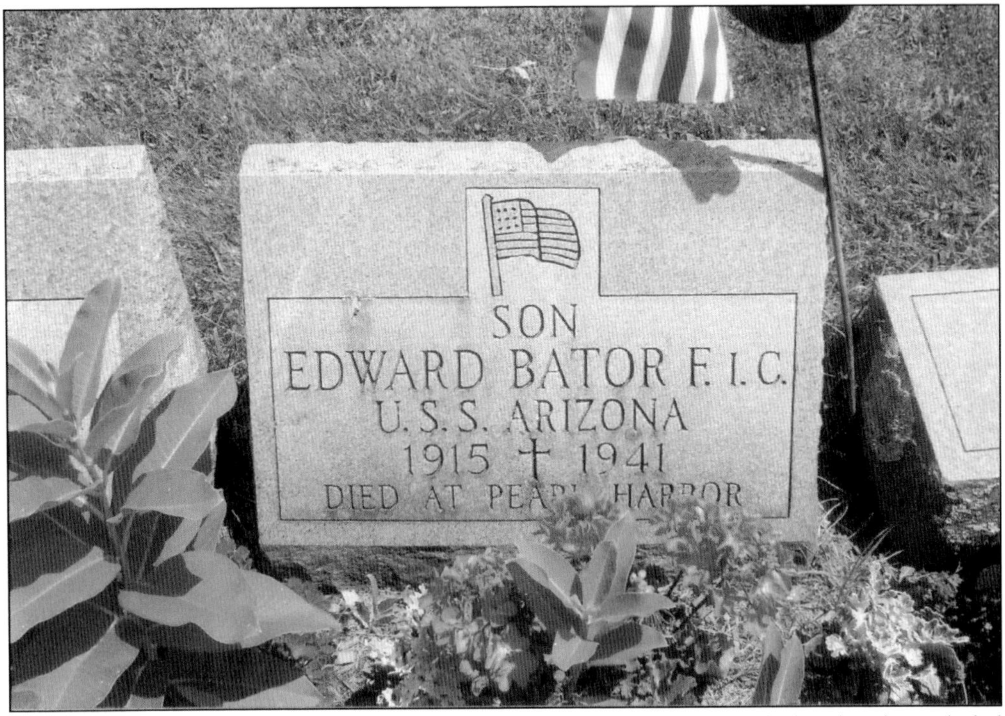

Eddie Bator was assigned to the battleship USS *Arizona* where he was killed when bombs exploded its magazine during the Japanese attack on Pearl Harbor. He was the first person from the village to perish during the war. Although his body remained entombed in the *Arizona*, his family erected this headstone in the Polish National Catholic cemetery in his honor. (Authors' collection.)

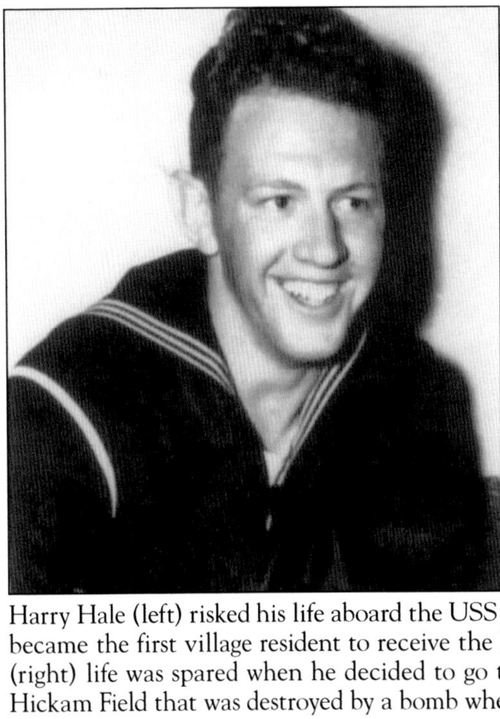

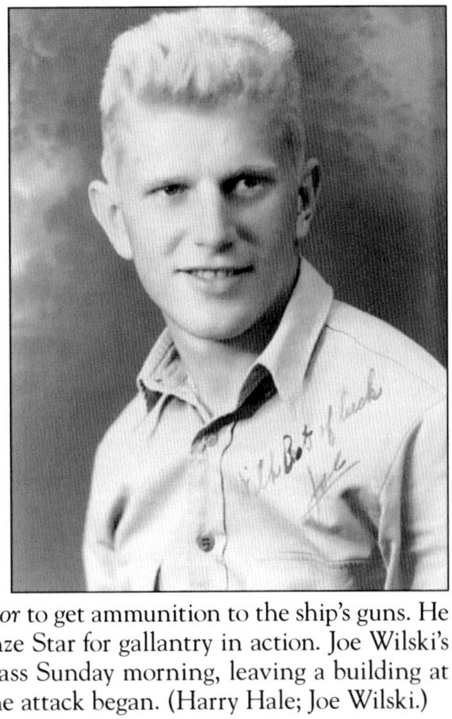

Harry Hale (left) risked his life aboard the USS *Trevor* to get ammunition to the ship's guns. He became the first village resident to receive the Bronze Star for gallantry in action. Joe Wilski's (right) life was spared when he decided to go to mass Sunday morning, leaving a building at Hickam Field that was destroyed by a bomb when the attack began. (Harry Hale; Joe Wilski.)

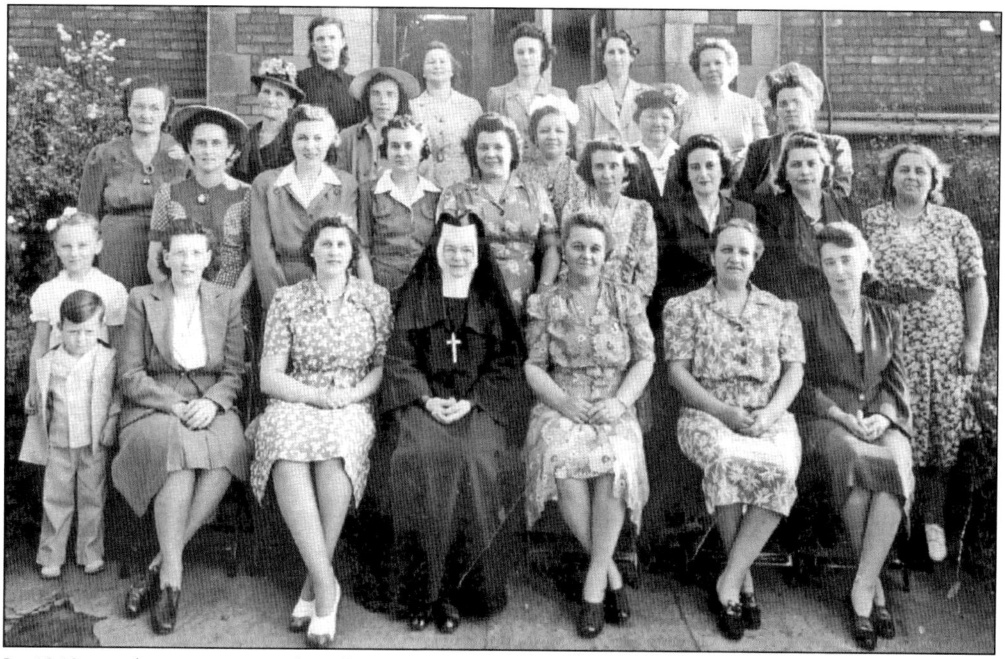

In 1943, on the suggestion of Stella Furgal, a Mother's Club was established at St. Mary's to raise funds to support the parish school. The first members, in addition to Furgal, were Sr. Mary Eymard, Anna Juchniewicz, Lillian Malecka, Helen Kobielski, Mary Kiersztyn, Matilda Stolarczyk, Jadwiga Wolkonowska, Wanda Krol, Sophie Bara, Helen Snizek, and Stella Marcinkiewicz. (NYMHS.)

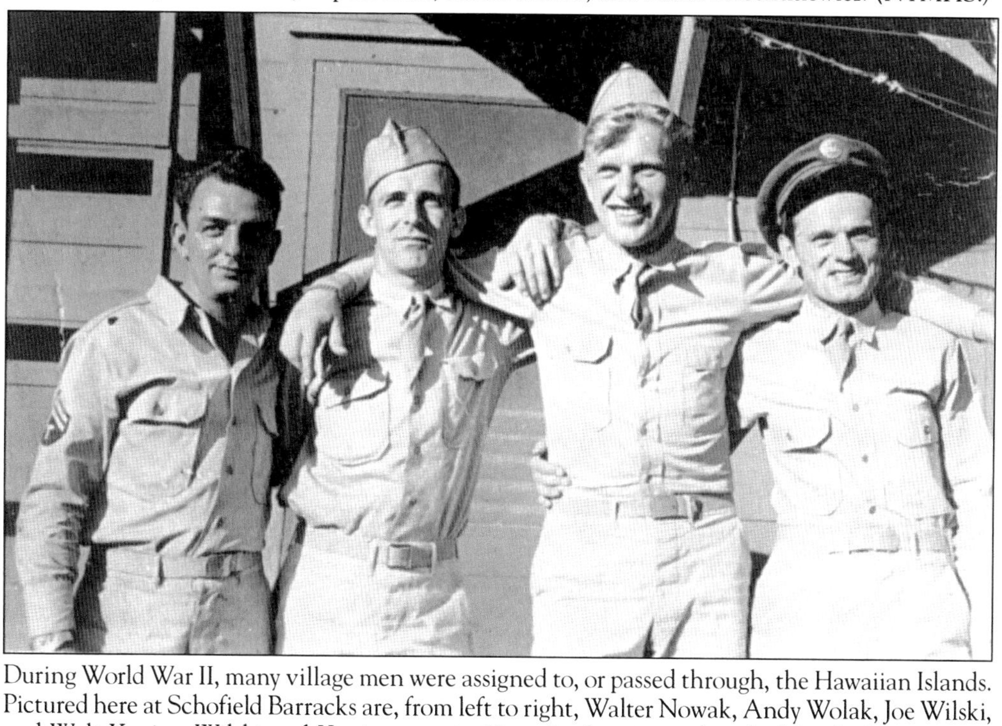

During World War II, many village men were assigned to, or passed through, the Hawaiian Islands. Pictured here at Schofield Barracks are, from left to right, Walter Nowak, Andy Wolak, Joe Wilski, and Walt Kupiec. Wilski and Kupiec were in Hawaii when it was bombed by the Japanese on December 7, 1941. All four survived the war to return home. (Joe Wilski.)

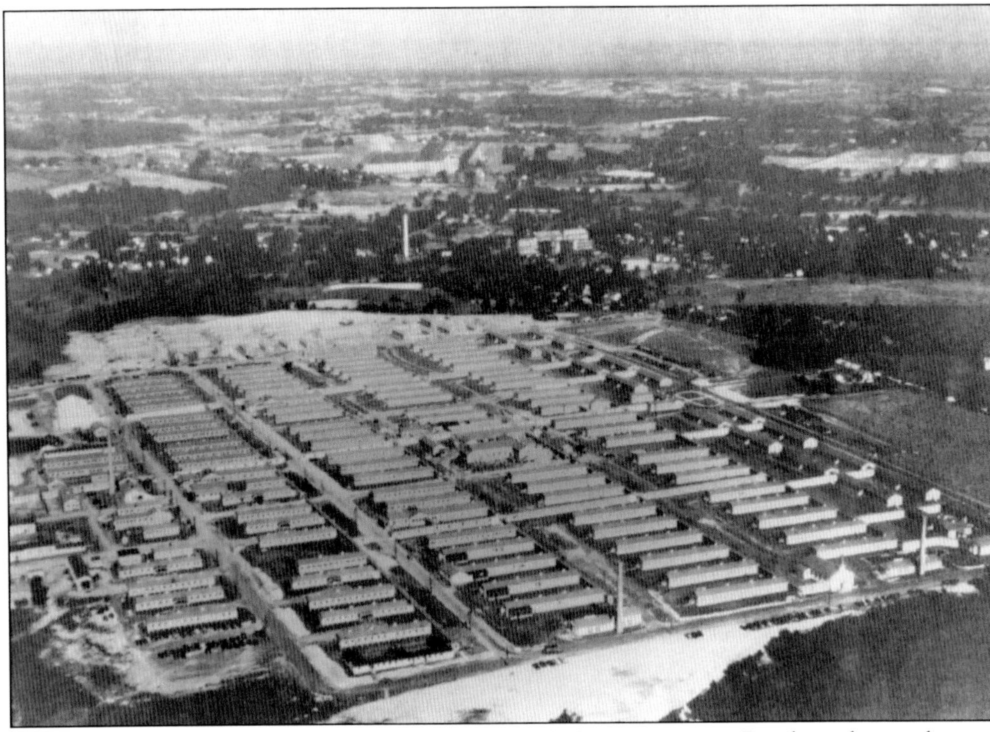

In 1942, the US Army opened the Rhoads Hospital along Burrstone Road on the outskirts of the village. Named for Col. Thomas L. Rhoads who served in the Spanish-American War, the $4.5-million, 170-acre facility included 126 buildings containing 1,500 beds. Before it closed in June 1946, the facility treated 25,277 patients. Part of the land later became the site for the new high school constructed in the 1950s. (UC.)

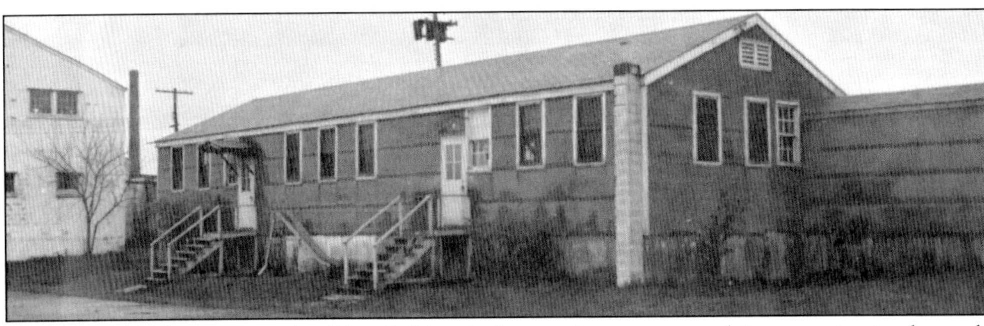

Buildings like this one on the Rhoads Hospital grounds were pressed into service in the early 1950s as classrooms for grades one through three when the Main Street school facility became overcrowded. There was no cafeteria or other amenities for the young children who attended classes there, only the bare walls, small desks, and an occasional chalkboard. (NYMPS.)

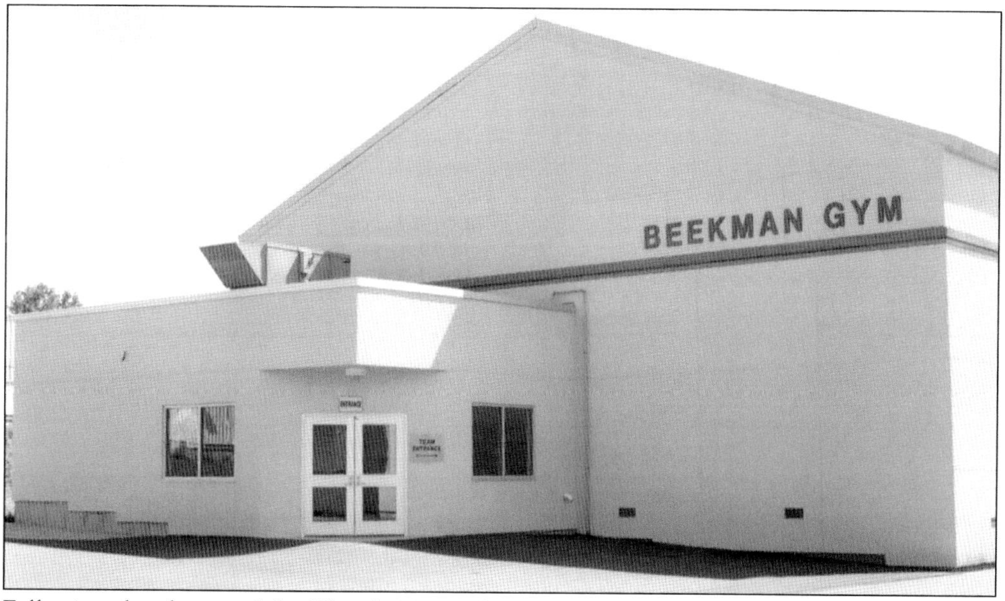

Following the closing of the Rhoads Hospital, its gymnasium was sold to the New York Mills Union Free School District for the price of $1. It became the location of high school athletic events and school intramural programs. It was eventually named Beekman Gymnasium in honor of longtime Mills fan John Beekman, who set an unequaled record for attendance at high school basketball contests. (Authors' collection.)

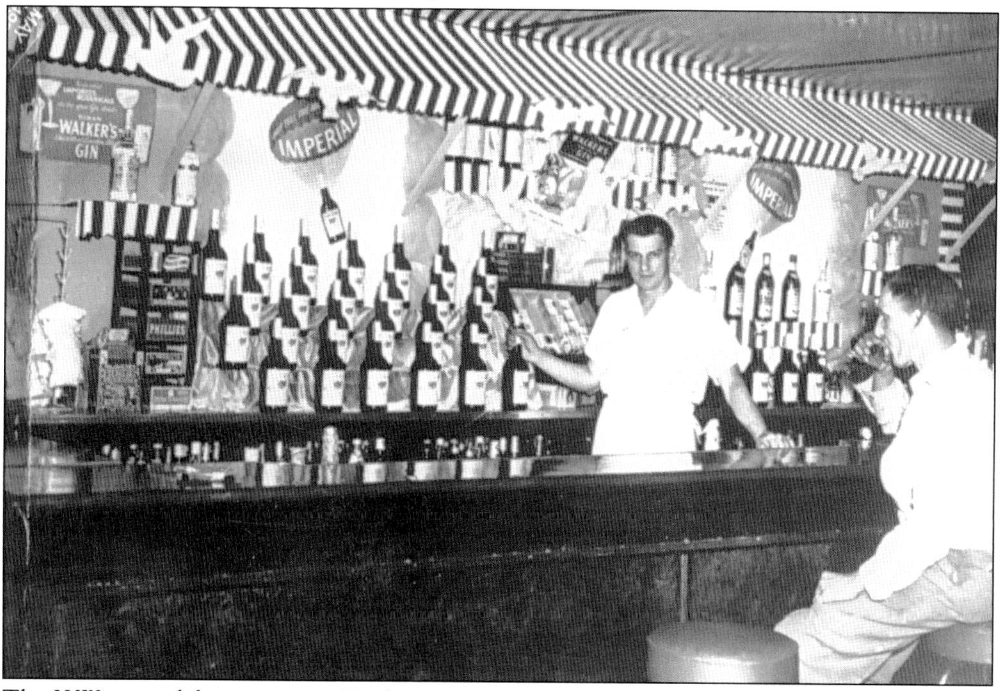

The K-W, named for its owners Hank Kupiec and Art Webb, was a popular tavern located on Clinton Street in the Upper Mills. During World War II, staff members at the Rhoads Hospital followed what they nicknamed the Burma Road to the K-W for lunches and after-work relaxation. The proprietors were well-known supporters of community projects. (Mildred Webb.)

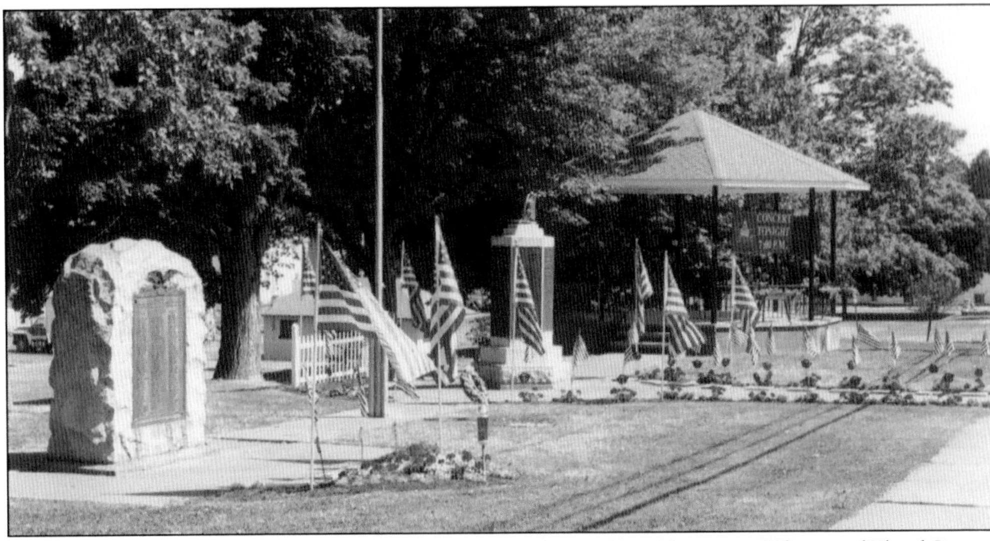

Following World War II, the A.D. Juilliard Company donated land between White and Floyd Streets adjoining Pulaski Park for the creation of Veterans Park in honor of those who served in the war. The World War I monument, originally located near the post office on Main Street, was moved to the new Veterans Park and a World War II monument erected nearby. (Authors' collection.)

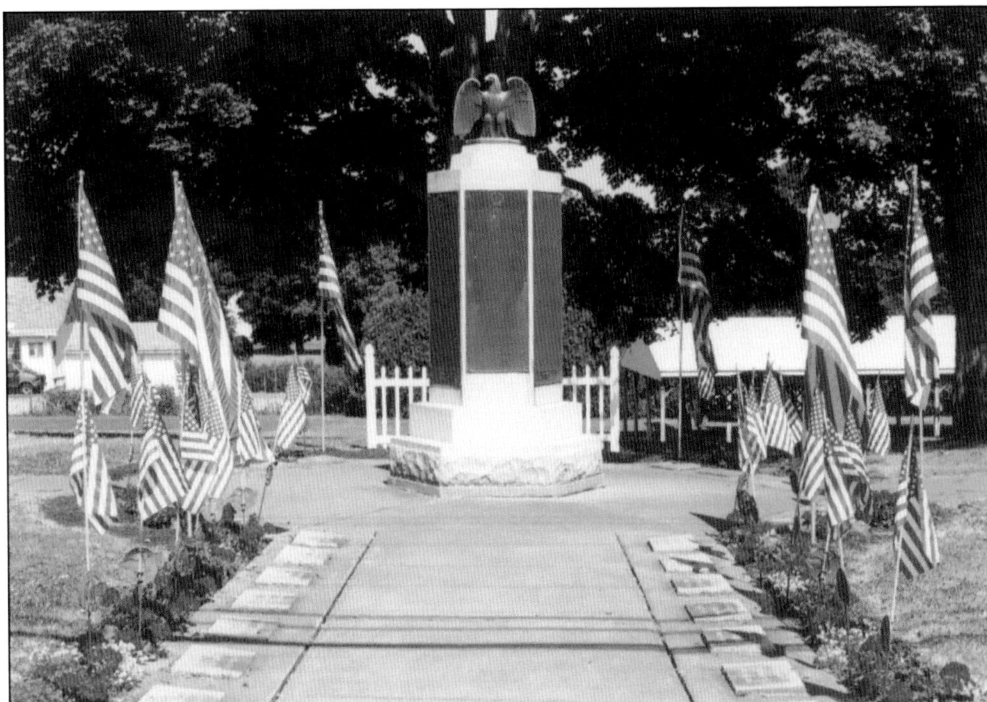

The World War II monument contains the names of those who served during the conflict. Twenty-three lost their lives: Barney Bator, Edward Bator, Richard Bilodeau, Martin Boris, Edward Dziedzic, Clayton Harrington, Clifford F. Harrington, Walter Jarosz Jr., Walter Kapalka, Napoleon Krawczyk, Longin Kurowski, Edward Labuz, Stanley Labuz, Barney Mikus, Stephen Misiaszek, Walter Nogas, Leroy Ouimette, Stephen Purdy, Gerald Spencer, Edwin Trzepacz, Peter Unczur, Walter Williams, and Leo Wojcik. (Authors' collection.)

Six

A Century Ends, a New Evolution Begins

The postwar years witnessed yet another major evolution in the community when the last of the textile mills closed. Without the constant hum of the machinery, the village became eerily quiet deepening the foreboding prevalent with the loss of jobs that came with the closing. While many outsiders predicted New York Mills' demise, it emerged from the dramatic shift even stronger than before. People found jobs in surrounding communities, broadening the base of employment and making it less dependent on a single industry. The textile community company town transitioned very quickly into a typical suburban bedroom community where people resided and enjoyed the camaraderie of friends and family but where few actually worked.

With the passing of the older generation, some family businesses closed, others opened, and eventually the emergence of large discount department stores and chain supermarkets eroded what was once a thriving middle class of independent entrepreneurs. The golf course enjoyed a new popularity, but older institutions such as National Park gradually faded from the scene. Once popular groups such as the Filharmonia Choir and the Polish National Alliance similarly began to see their numbers dwindle as the number and variety of organizations within easy driving range competed for membership. Without the commonality of the workplace, people gradually began to affiliate more with unions where they worked and with social clubs including those with whom they worked on a day-to-day basis.

But village identity remained strong. Old Home Days brought people from all over the country to an annual celebration. When it eventually lapsed, the new Bell Festival took its place as an annual event for community celebration and the return of those whose lives had taken them elsewhere. Community improvements continued, not the least of which was a badly needed new school. A modern library appeared, along with a new firehouse and municipal building. Life had changed once again, but the community continued to adapt with each new evolution.

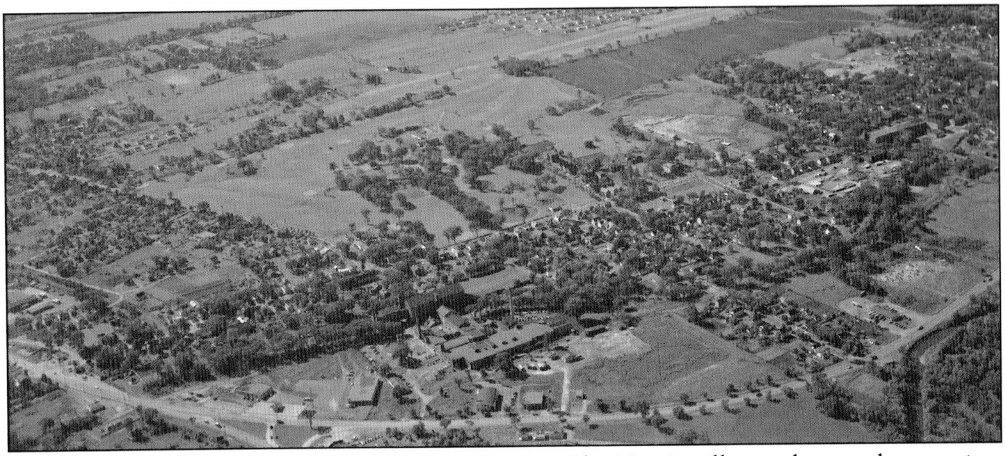

This aerial photograph was taken in the mid-1950s. The No. 1 mill complex can be seen just below the center and the No. 2 at the upper right. The curving road at the bottom is Commercial Drive before the interchange into Whitesboro was completed. To the right of the large open area in the center is St. Mary's Street with Union Hall and St. Mary Our Lady of Czestochowa Church. (Authors' collection.)

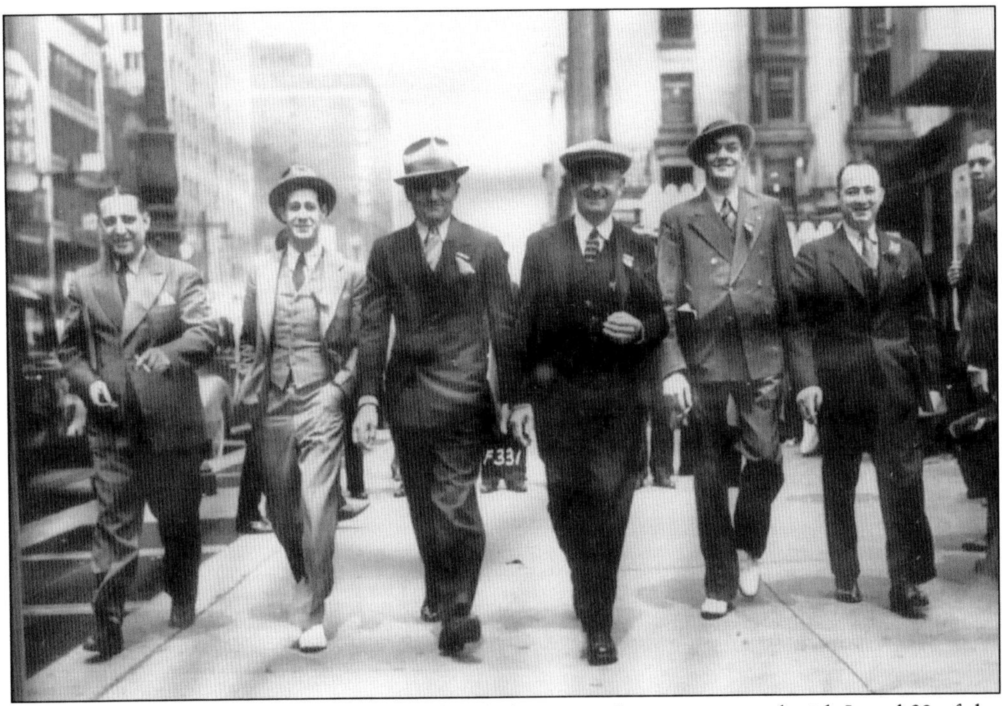

Labor strikes continued between 1930 and 1950, but many became internal with Local 32 of the Congress of Industrial Organizations vying for worker support with Local 753 of the American Federation of Labor. The infighting often produced friction between old friends and even within families. In the end, Local 753 prevailed. Village labor leaders are seen here in New York City. (Authors' collection.)

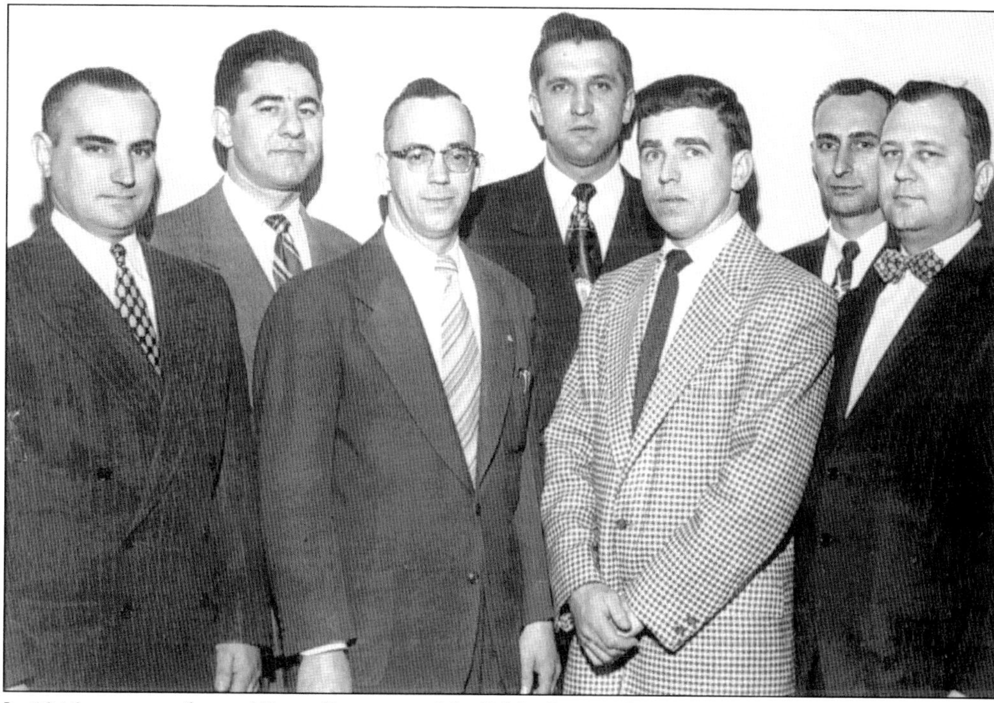

In 1946, veterans formed Bator Post named for Eddie Bator who was killed on the USS *Arizona* at Pearl Harbor. Its first president was Michael Pasternak. In 1953, its officers were, (from left to right) Frank Potrzeba, Stephen Pawlinga, Edward Szwedo, Joseph Palik, Stephen Zurek (commander), Stanley Potrzeba, and Stanley Maliszewski. In 1951, it became Post 21 of the Polish Legion of American Veterans. (Stephen Zurek.)

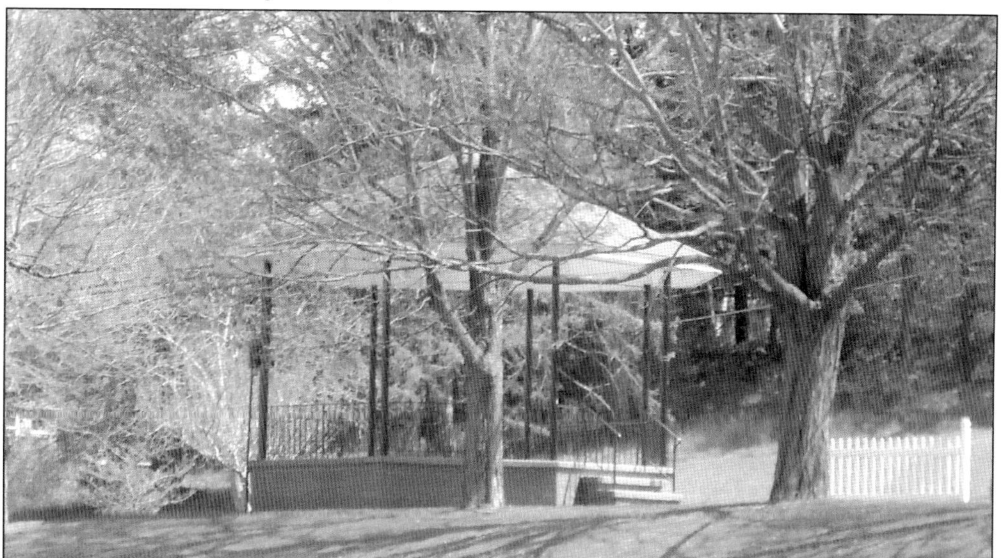

This gazebo was constructed and maintained for events in Veterans Park in 1980 by Bator Post 21 of the Polish Legion of American Veterans, the New York Mills Lions Club, and the Village of New York Mills. On Memorial Day and other civic occasions, the gazebo is used as a rostrum for speakers and guests. (Authors' collection.)

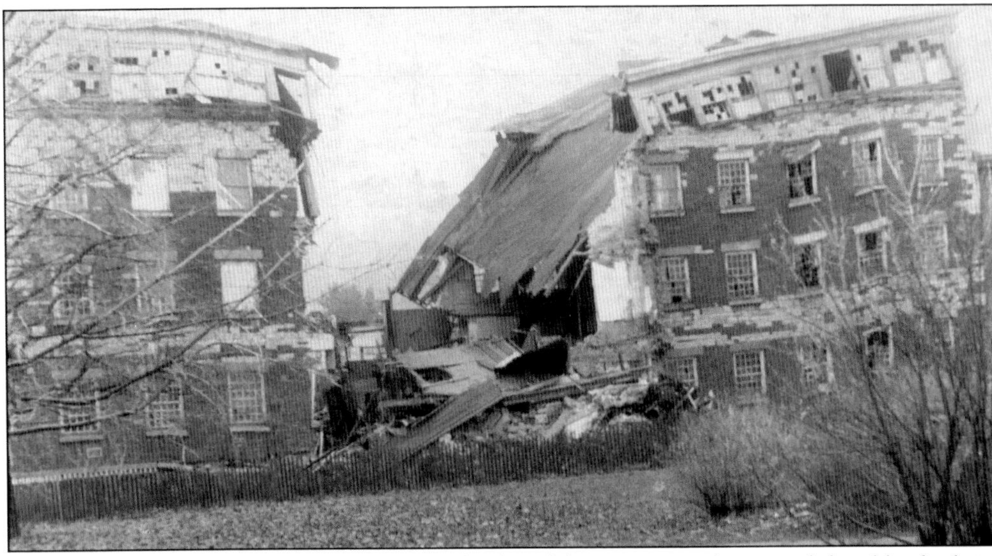

After the war, textile production moved south where labor costs were lower and the older facilities could be replaced with modern plants to increase efficiency. The last active mill was the No. 2, which closed in 1951. In 1976, the Middle Mill Historic District was added to the National Register of Historic Places, but before renovations could take place the center portion of the structure collapsed on November 8, 1977. (Authors' collection.)

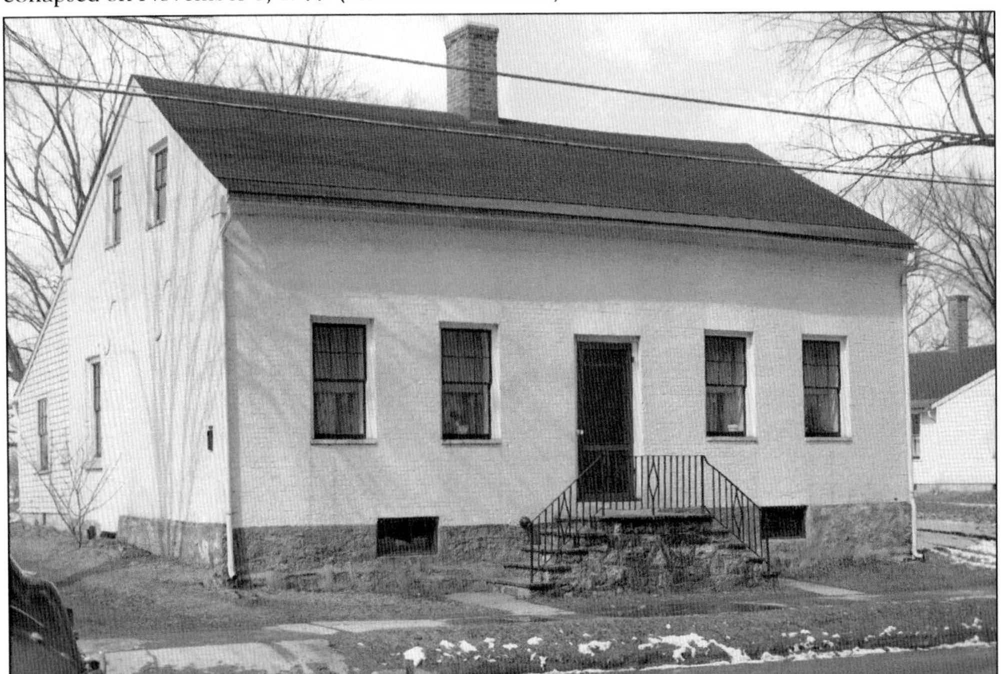

When A.D. Juilliard left the village, it sold its company-owned housing, giving current residents the first option to buy. Prices were set lower than the prevailing market rate to allow people to keep their homes. Since most were two-family homes, this often led to humorous circumstances when the owner of half of the house chose to paint it a very different color than the other half. (Authors' collection.)

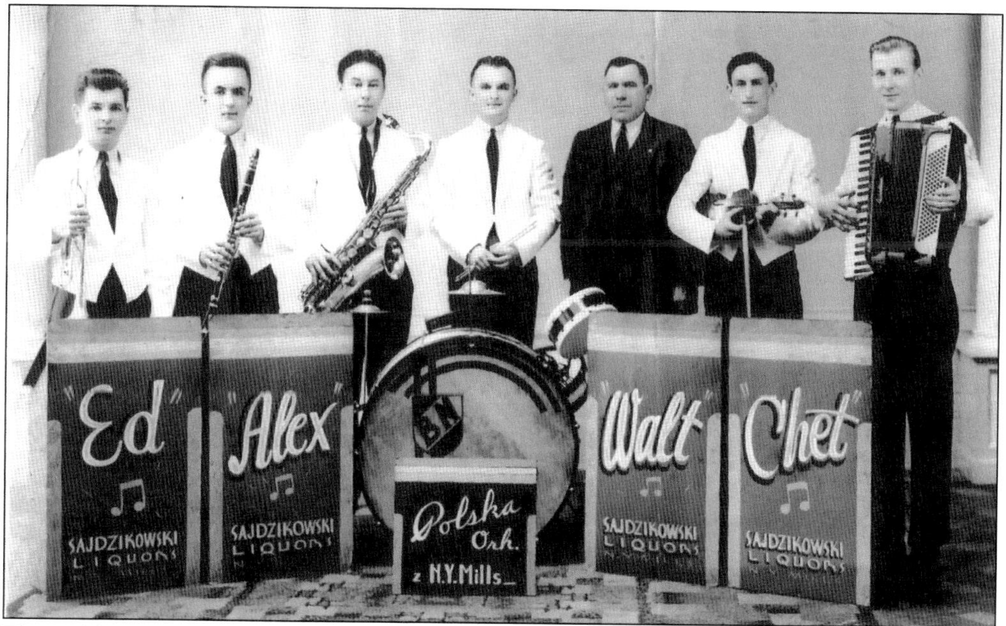

The Polska Orchestra, sponsored by Sajdzikowski Liquors is a good example of the changing nature of Polish American bands. A trumpet and large drum set have replaced most of the strings found in Polish groups and those established by Polish Americans of the immigrant generation. This change began with the second generation in the interwar period and is clearly visible here. (NYMHS.)

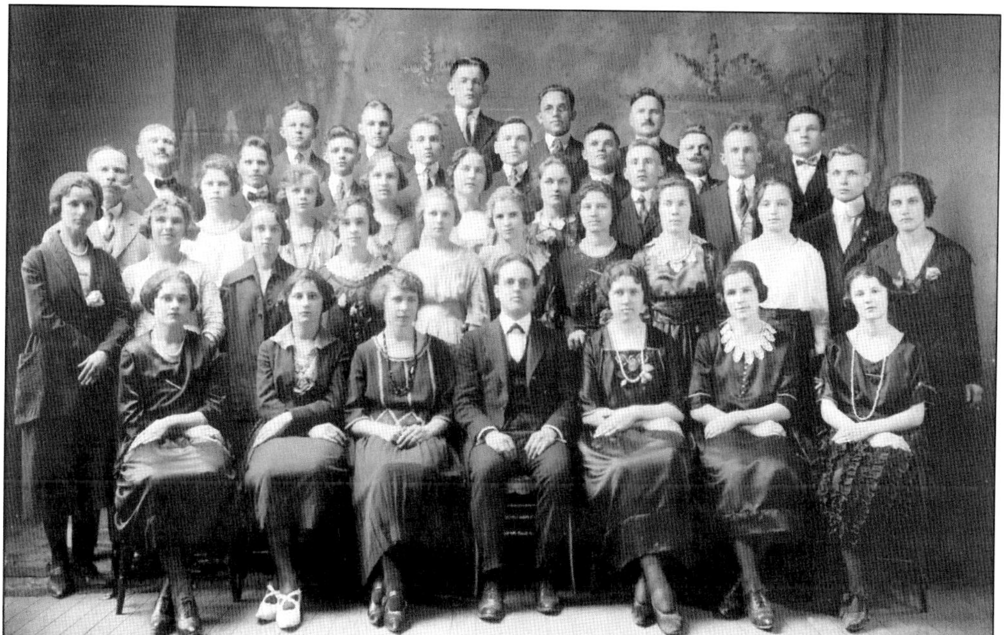

Musical director Joseph F. Kangiser, seated in the center, founded the Halka Ladies Choir in 1936. It took first place in a statewide competition during its first year. Kangiser also directed the Filharmonia Choir, a mixed gender group that the Utica *Observer-Dispatch* called "one of the most successful and better known choral groups in the area." (Authors' collection.)

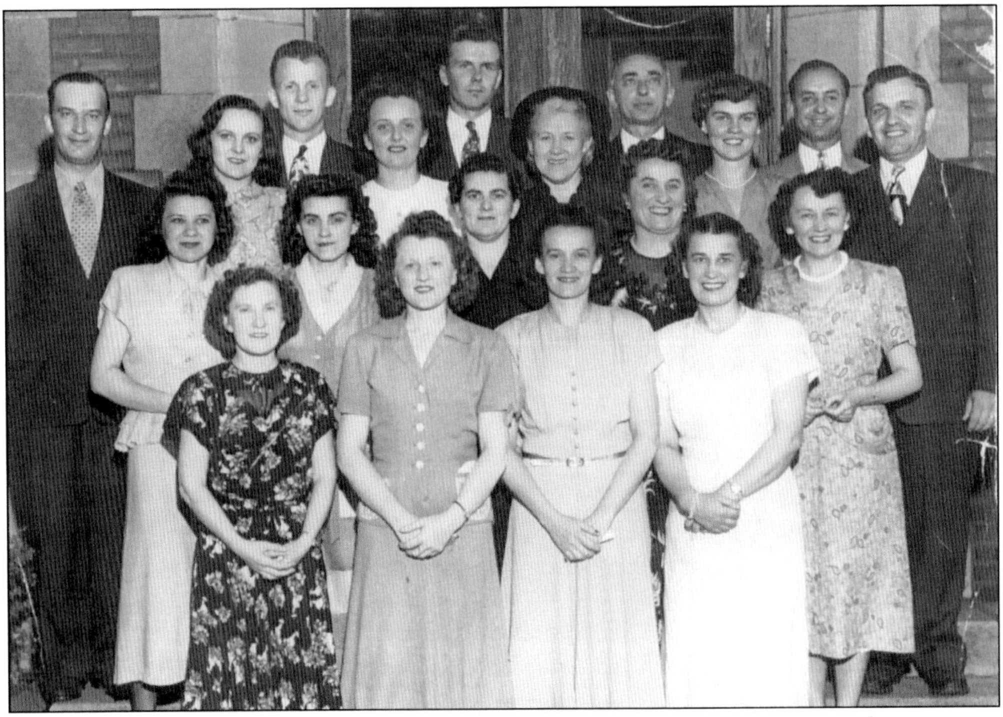

The Filharmonia Choir that won the Hlond Trophy as the best group in an international competition in 1947 included, from left to right, (first row) Stephanie Chrabaszcz, Jessica Adamczyk, Aniela Pikul, and Jane Kobielski; (second row) Helen Dziedzej, Aniela Krol, Helena Smolka, Sophie Soja, and Wladyslawa Miga; (third row) John Kobielski, Helen Zima, Stan Zima, Marian Kaczor, Emil Labuz, Stephanie Nowicki, Leo Soja, Jeannette Furgal, Walt Bogart, and Joe Kaczor. (Labuz.)

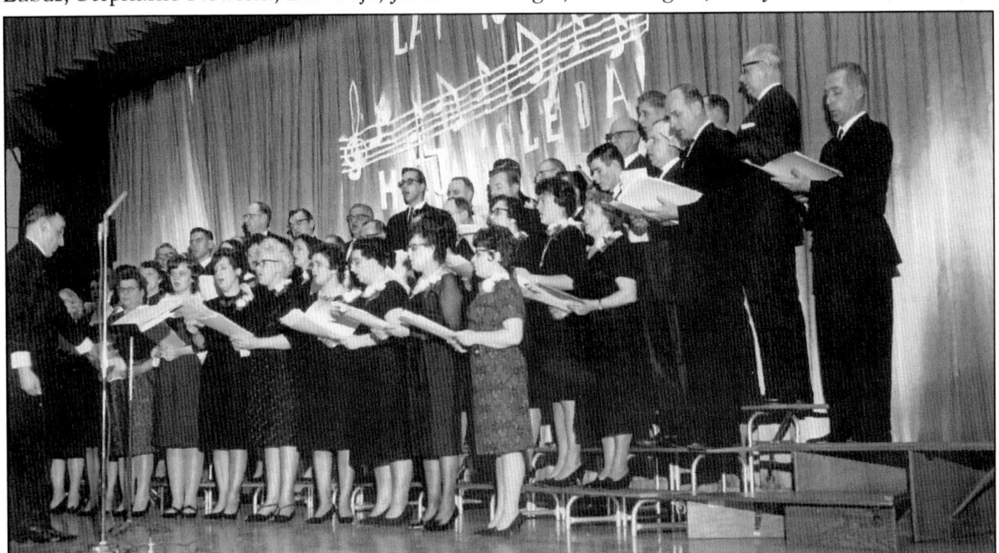

The Filharmonia Choir performs here in the high school auditorium on the millennium of the Polish nation in 1966. Originally founded as St. Cecelia's Choir in 1911, the group changed its name to Filharmonia in 1921, joined the Polish Singers Alliance as Group 120, and won an unprecedented 13 first-place titles in district and national competitions. (OCHS.)

Al Pezdek, pictured here, owned a gas and service station in the center of the village for decades. A prominent supporter of civic activities, he managed the Islanders, a local baseball team, and was an early backer of the New York Mills Little League including sponsoring one of the teams. (Babiarz-Pytko.)

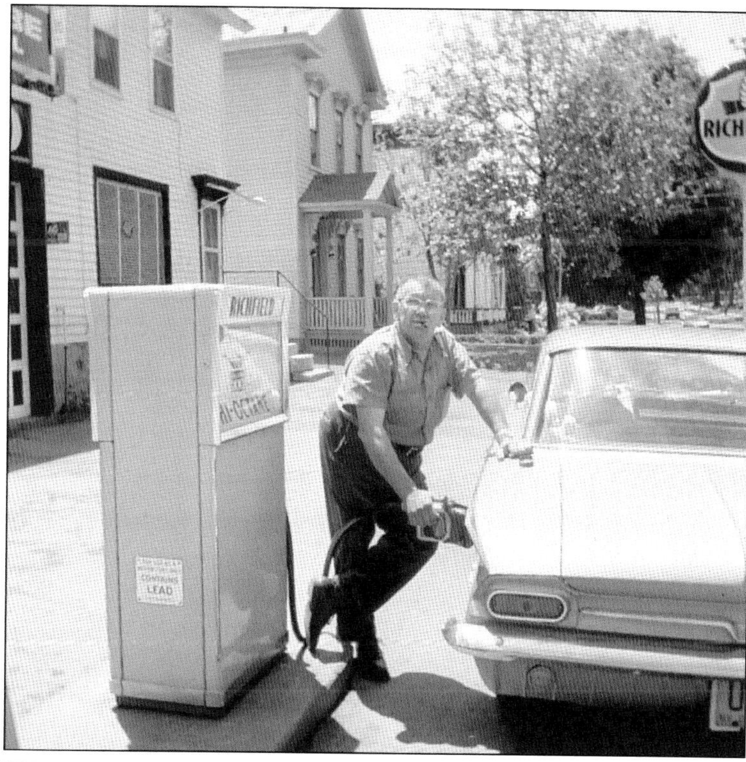

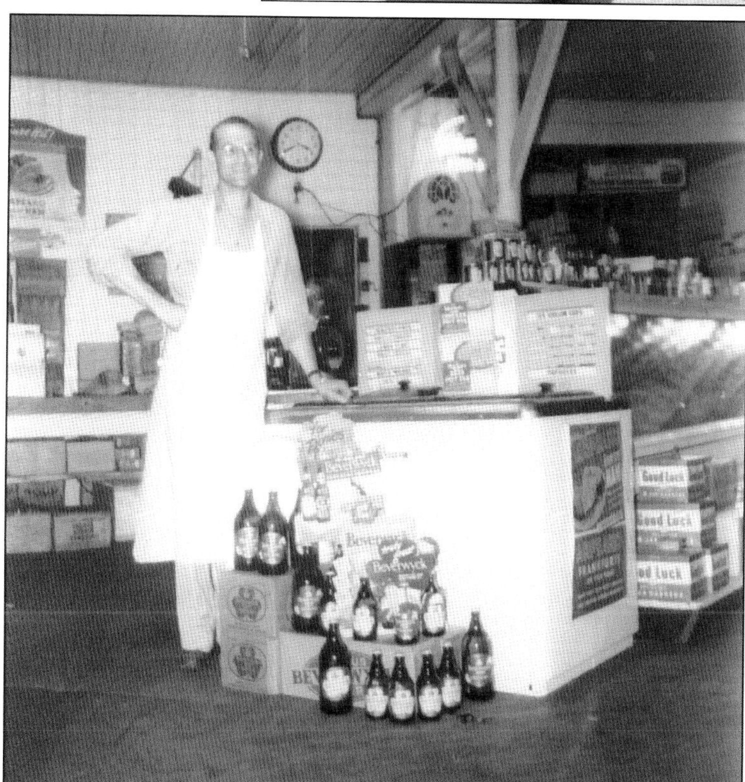

Ed Kozak and his wife, Olga, owned Kozak's market on the corner of Clinton and New Hartford Streets where they provided the neighborhood with fresh vegetables, dairy products, meats, and other grocery items. Since they owned their own home across the street, the second floor over the market was rented to boarders. (Unczur.)

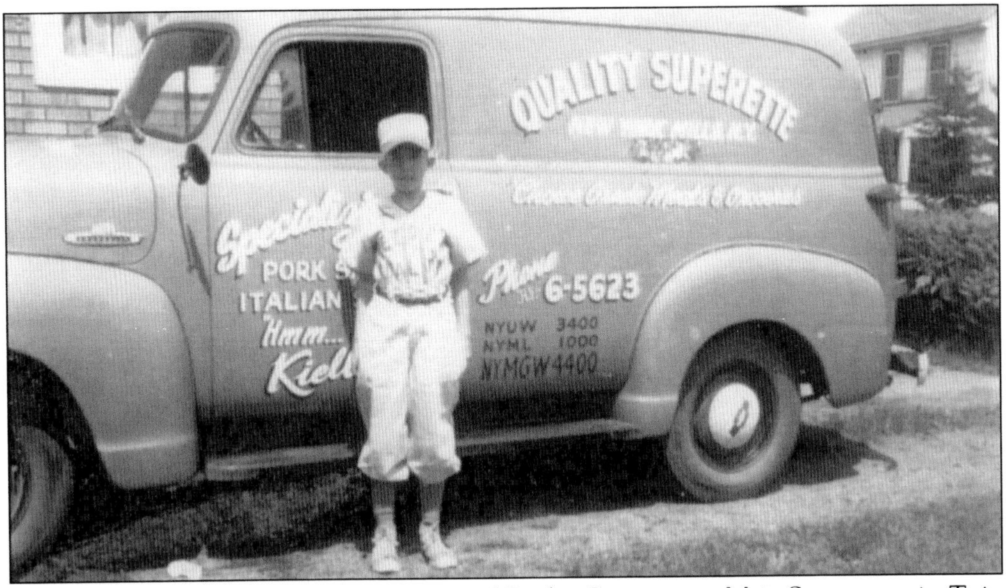

Barney and Chester Dziedzic were co-owners of Quality Superette on Main Street opposite Twin Ponds. The truck was used for delivering groceries, a personalized service that retained a large base of customers. Standing by the truck is Bobby Dziedzic, Barney's son, in the uniform of the Little League team the Superette sponsored. (OCHS.)

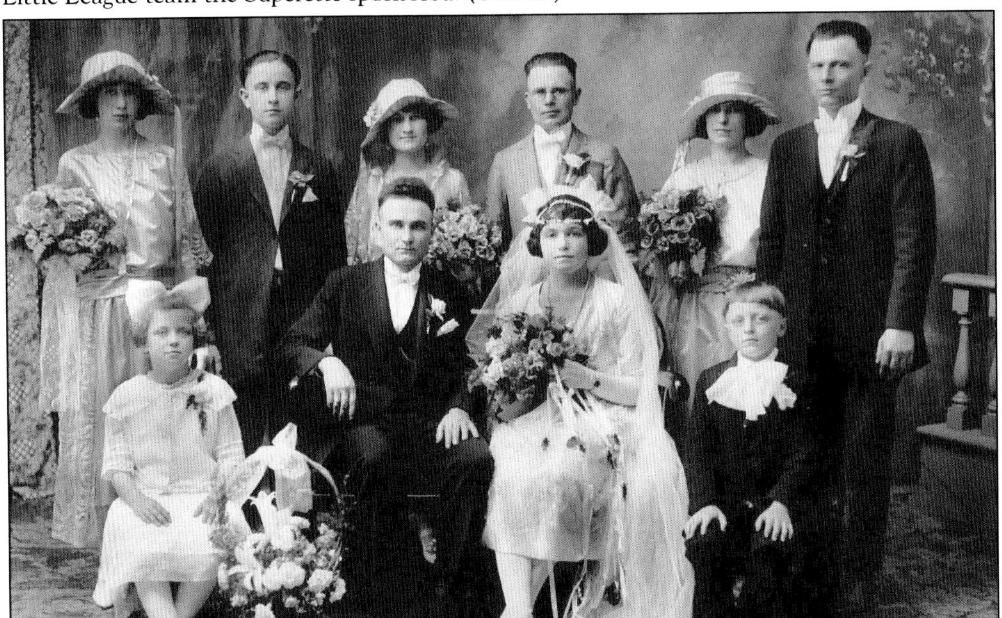

Weddings were major social events not only for the couple being married but also for the whole neighborhood and extended community. Steven and Mary Semeniuk's nuptials illustrate the elegant attire, elaborate flower arrangements, and formal portrait required on these occasions. It was not uncommon for the reception following the ceremonies to last all weekend until the participants had to go to work on Monday morning. (C. Pula)

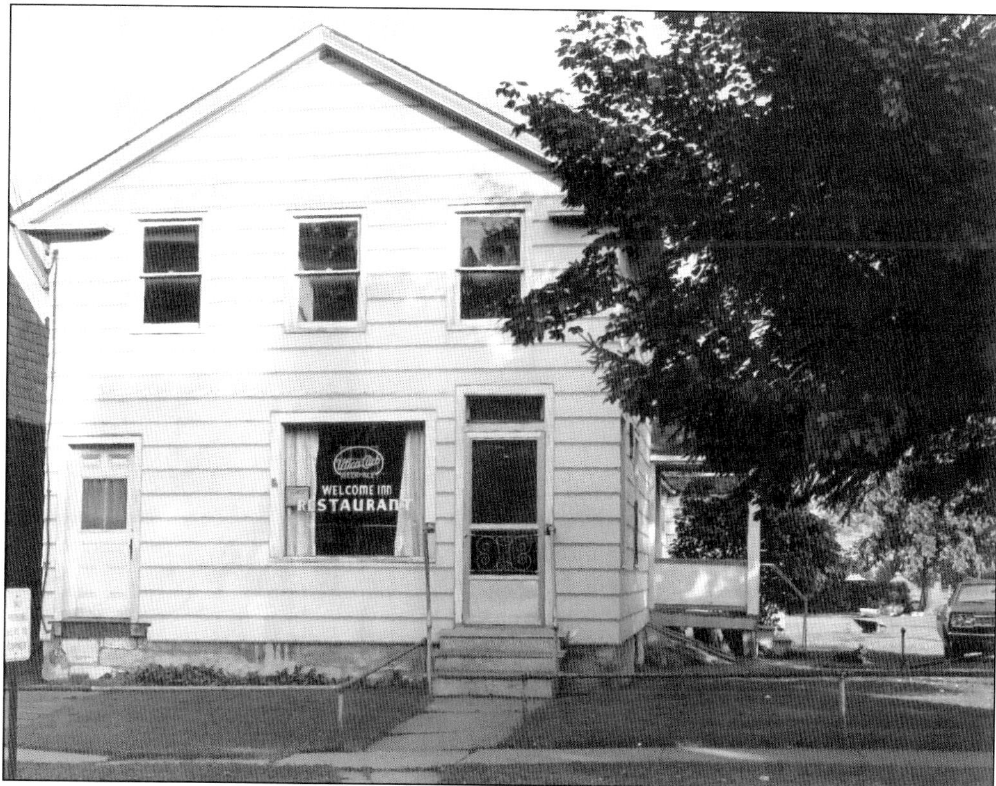

Located in the Lower Mills, the Welcome Inn was typical of many businesses in the postwar years. As automobiles and suburban shopping areas led to the demise of most family-owned local grocery stores and butcher shops that had dotted the community in earlier years, taverns survived by providing a handy place for people to gather and relax near their homes. (Authors' collection.)

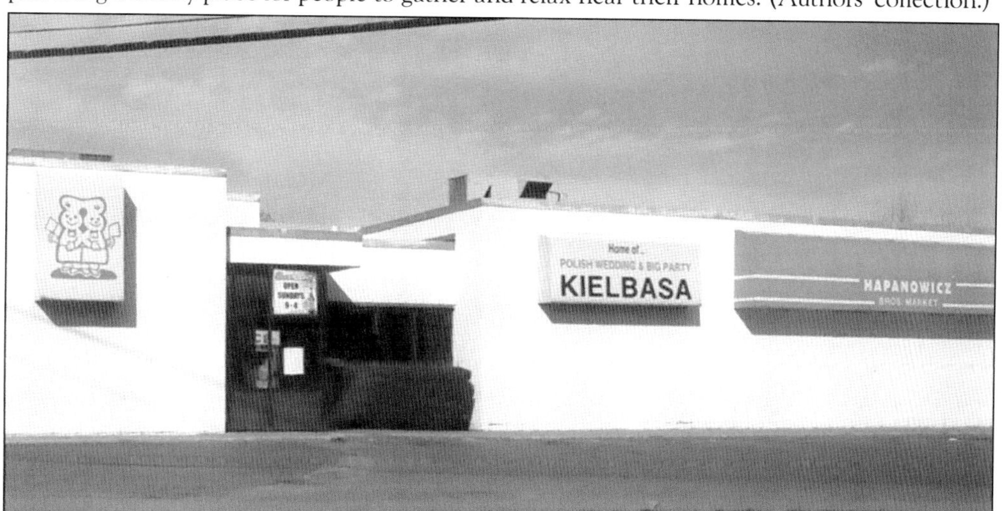

For a century, the population of the Upper Mills was served by the typical corner grocery store: first McCarthy's, then Kozak's, then Chester's Superette. Eventually, the building that successively housed these businesses was torn down and replaced by Hapanowicz Brothers Market, a new, modern facility specializing in quality meats. (Authors' collection.)

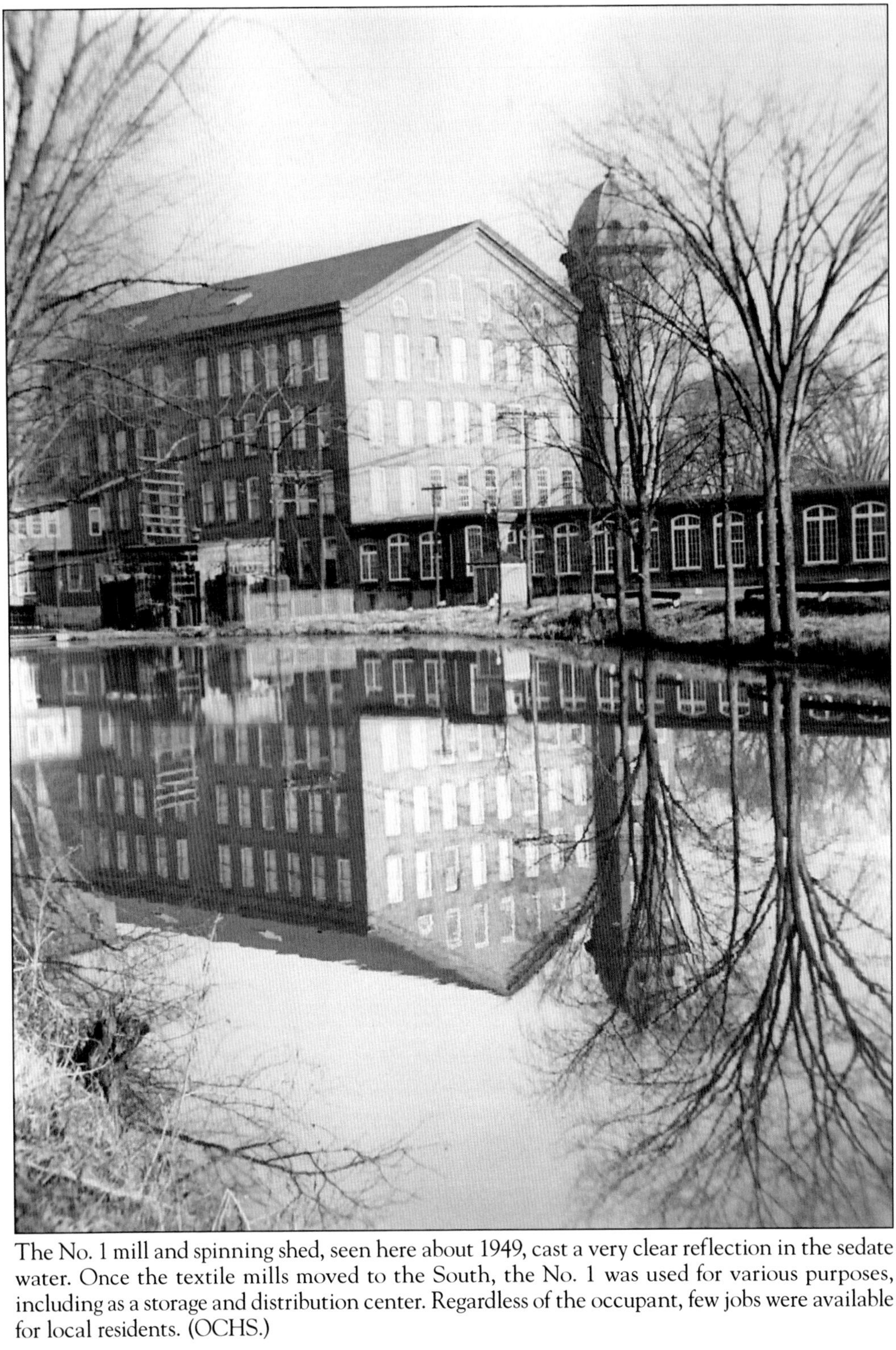

The No. 1 mill and spinning shed, seen here about 1949, cast a very clear reflection in the sedate water. Once the textile mills moved to the South, the No. 1 was used for various purposes, including as a storage and distribution center. Regardless of the occupant, few jobs were available for local residents. (OCHS.)

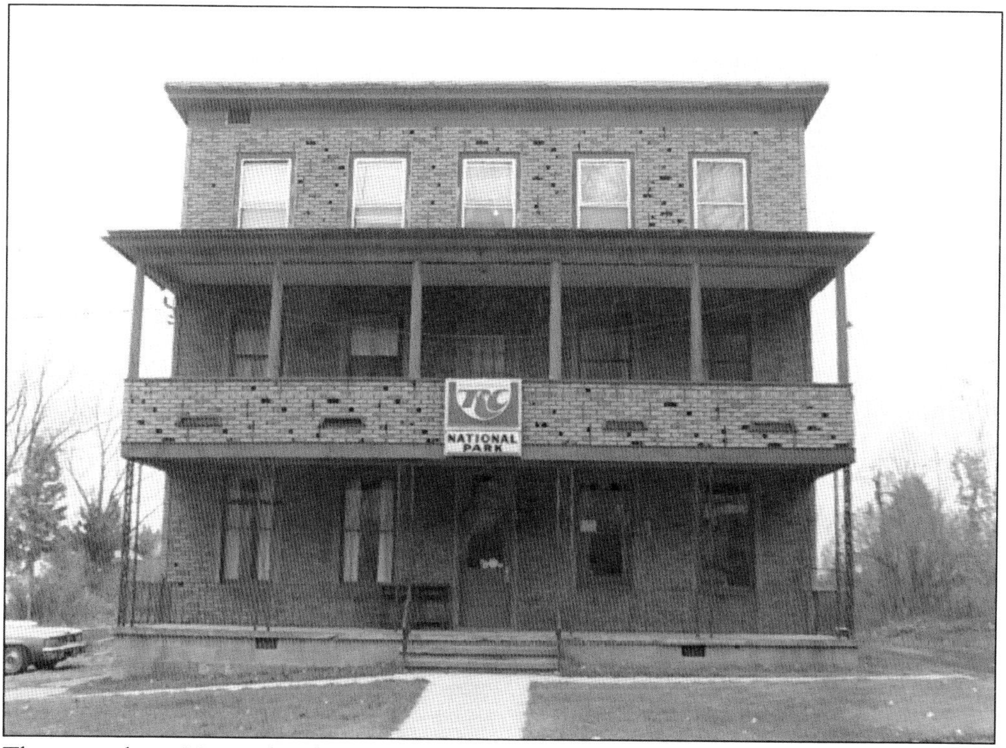

The once vibrant National Park on Porter Street—the scene of picnics, celebrations, and weekend festivities for so many years—eventually lost out to newer amusements and social events outside the village. Shown here well past its glory days, the building was eventually destroyed by fire in December 1982, punctuating the final ending to that era of community life. (OCHS.)

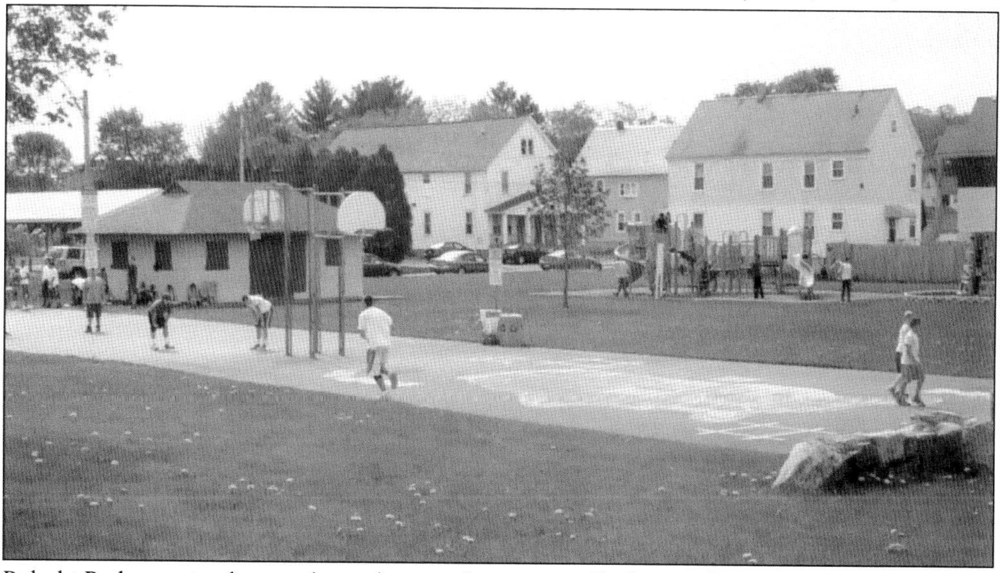

Pulaski Park remained a popular gathering place, especially for people with young children who enjoyed the playground, and for school-aged children who played basketball or just enjoyed an opportunity to "hang out" and socialize. On special occasions such as the annual Bell Festival, the park became a scene of general community gathering. (Authors' collection.)

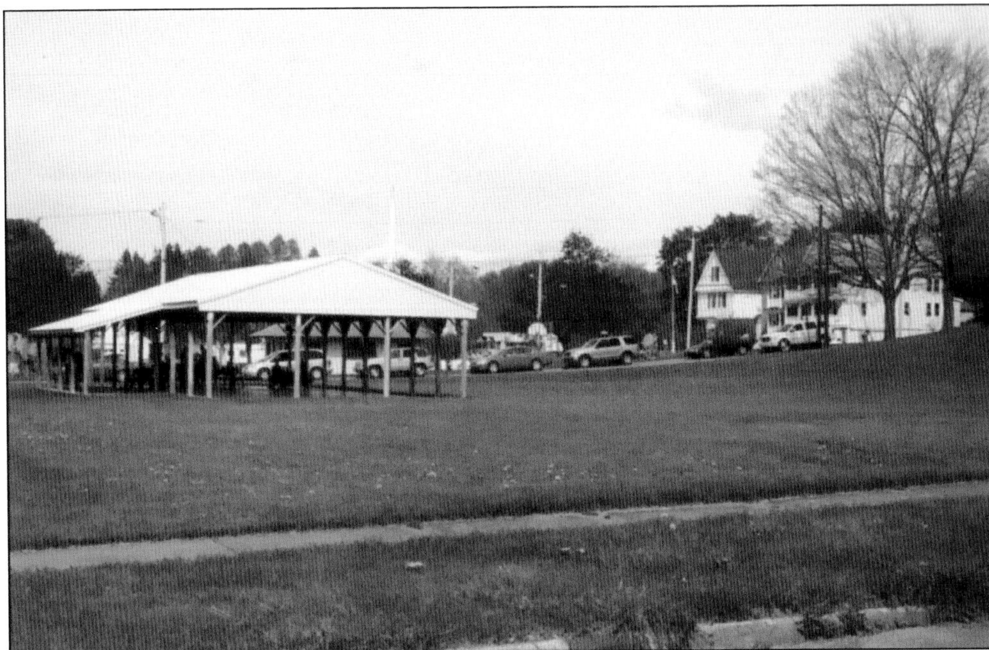

Established after World War II, Veterans Park to a certain extent replaced National Park as the community-gathering place. It hosted summer band concerts, fundraising activities, and other events. The picnic pavilion was used for community and family gatherings, as well as larger events such as the Bell Festival. (Authors' collection.)

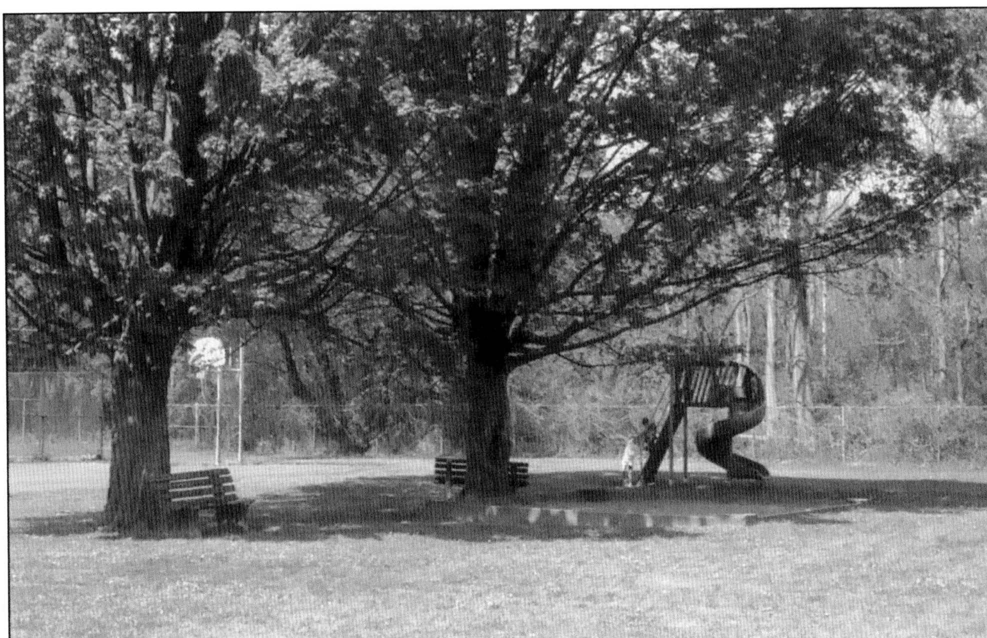

John Pietryka Park, located on land donated by the A.D. Juilliard Company between Clinton and Henderson Streets in the Upper Mills, was named for a longtime mayor who did much to improve village life. It contains the Fred Moses Tennis Center, playgrounds, and the New York Mills Little League field. (Authors' collection.)

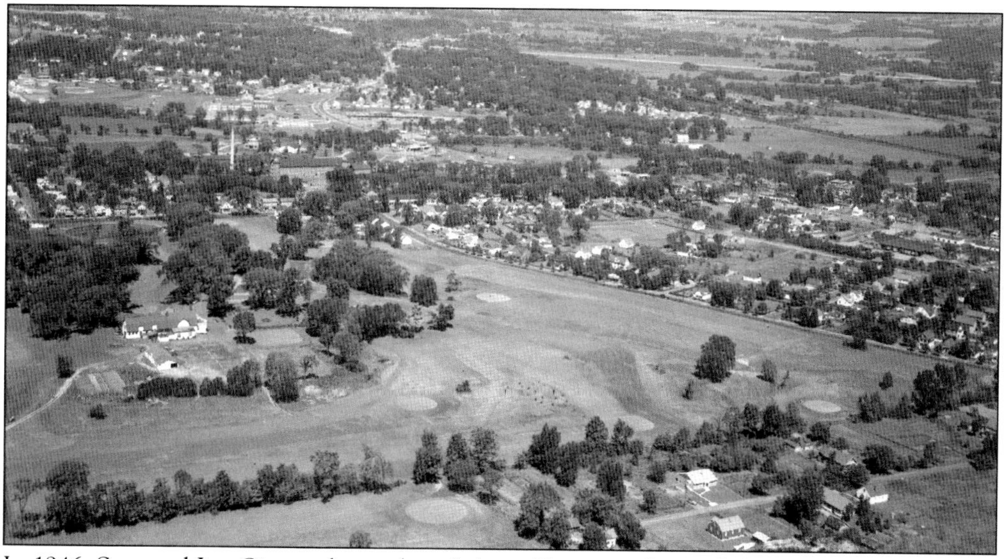

In 1946, Sam and Jim Girmonde purchased the golf course, changed the name to Twin Ponds, and began hosting local and regional golf tournaments including the annual Jenny Open. The home of several local leagues, it also gave residents an opportunity for employment and leisure activities. (Authors' collection.)

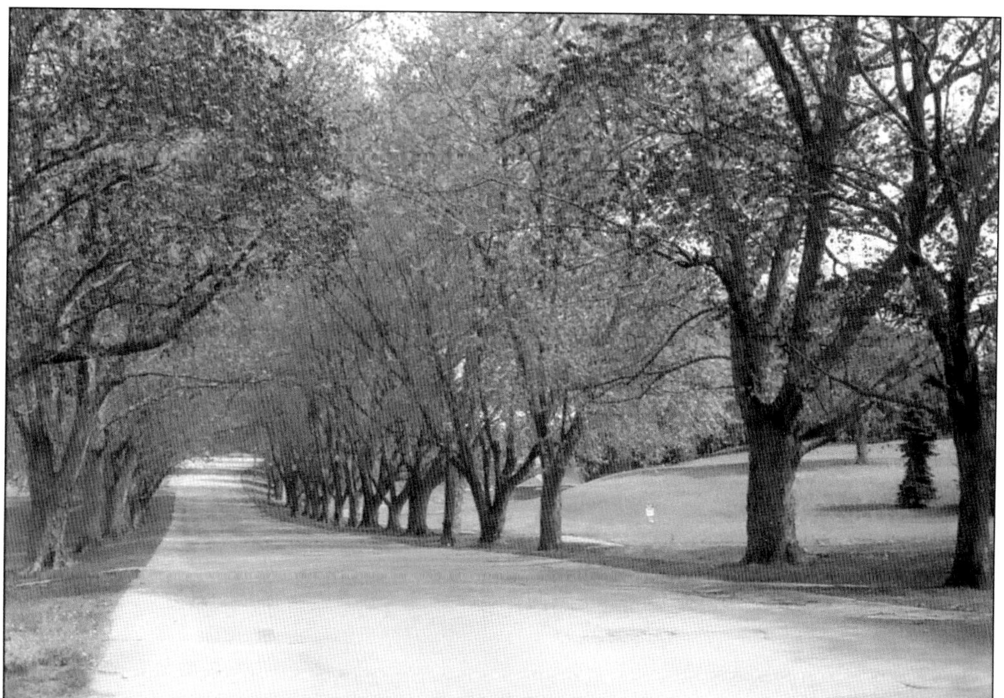

Twin Ponds, named for the water pools on each side of the entrance, produced prominent golfers including Ronnie Knych who won the Indiana State Championship five times, Matthew "Larry" Bartosek, the 1956 PGA Club Professional champion, Marty Furgal, the Western Open winner who played on the US Ryder Cup team, and Ted Krol, who won several PGA events and also played on the US Ryder Cup team. (Authors' collection.)

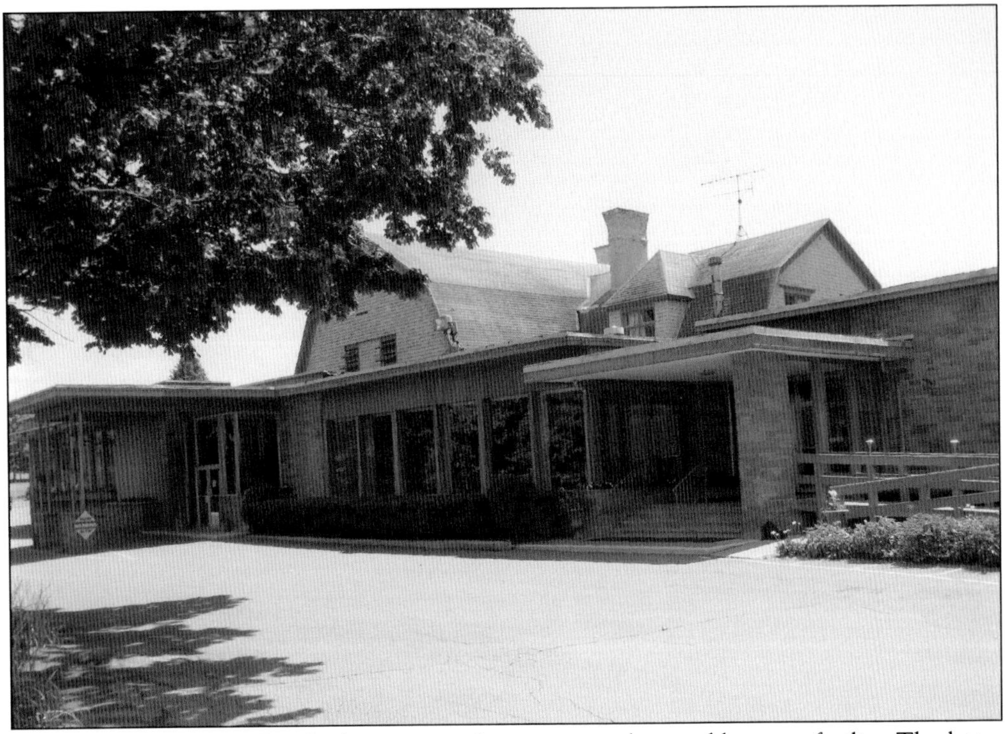

The clubhouse at Twin Ponds also contained a restaurant, bar, and banquet facility. The latter hosted wedding receptions, dinners, and other social events by groups from all over Central New York. Originally a barn on the Campbell property, over the years, it was refurbished and expanded with new construction to accommodate a growing list of services. The roof structure of the original barn is still visible. (Authors' collection.)

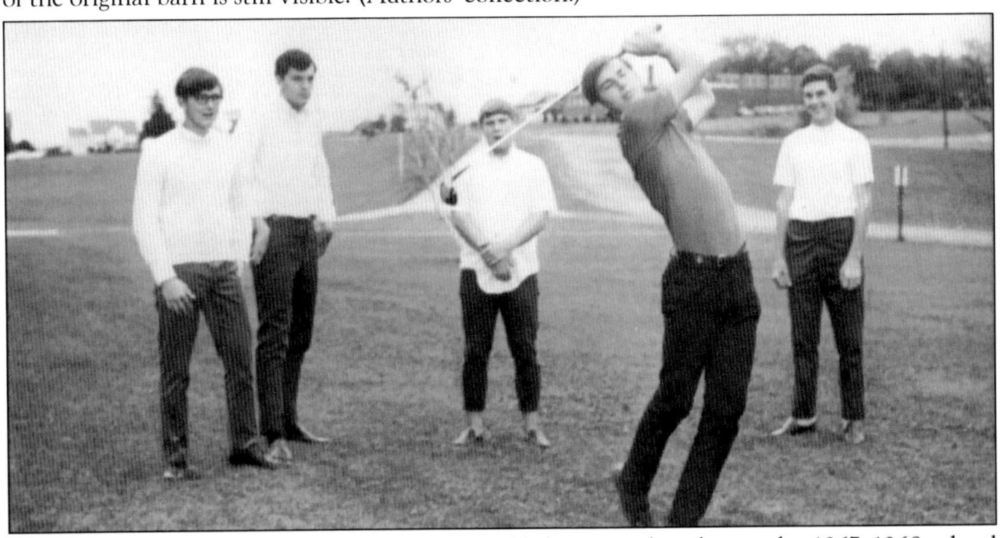

Larry Kupiec, winner of the New York State golf championship during the 1967–1968 school year, practices his swing in front of fellow teammates (from left to right) Jeff Gorczynski, John Zwijacz, Richard Furmanski, and Ron Czyzycki. The high school golf team was a perennial challenger in league and state matches, due in large part to the availability of the Twin Ponds facility. (C. Pula.)

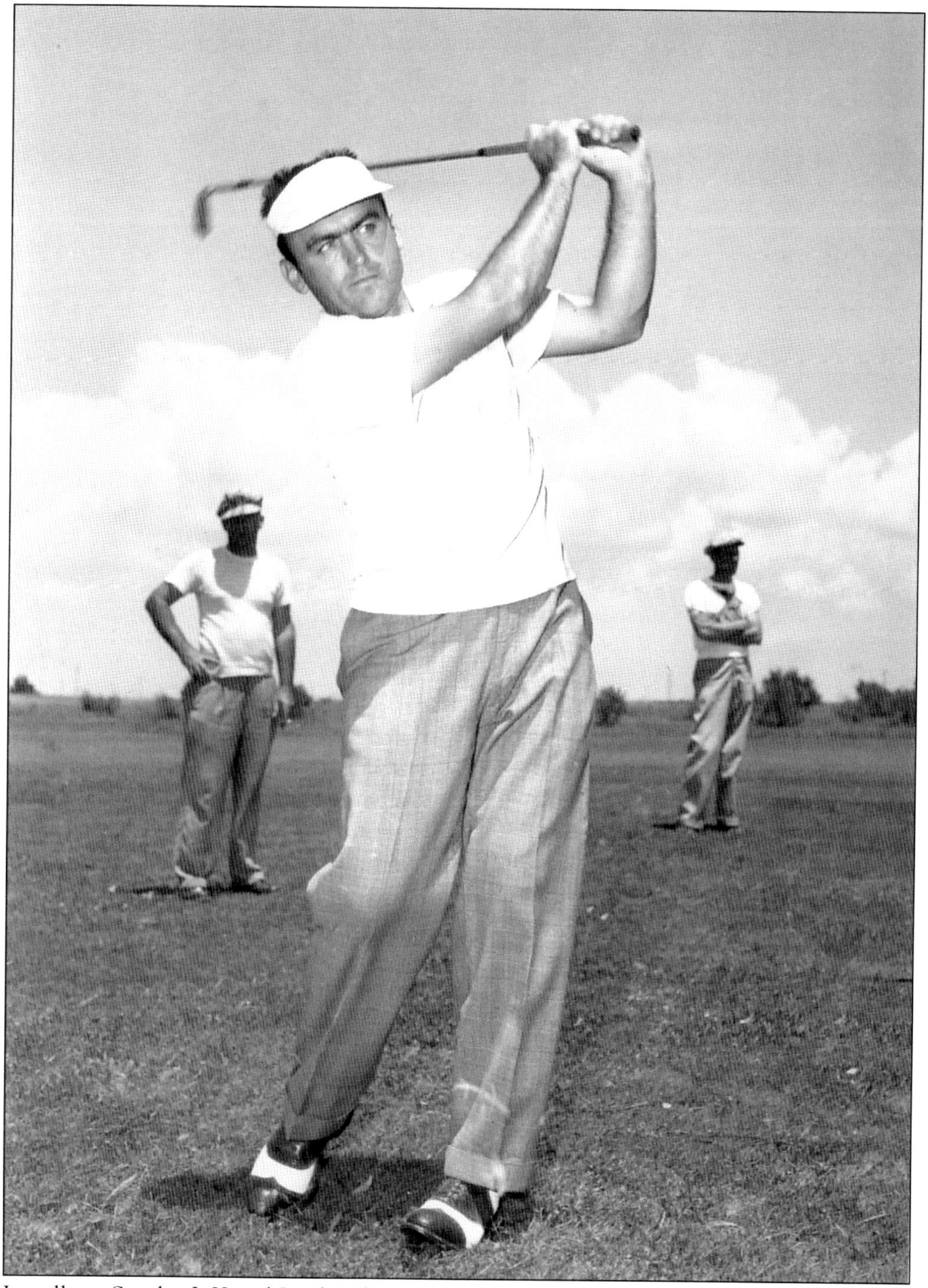

In college, Stanley J. Kowal Jr. played on the Michigan State University golf team in the Big Ten Conference. He won the Utica City Amateur Championship in 1962 and was a multiple winner of the Jenny Open. Like so many others, he began his venture into golf by caddying at Twin Ponds. (Kim Kowal.)

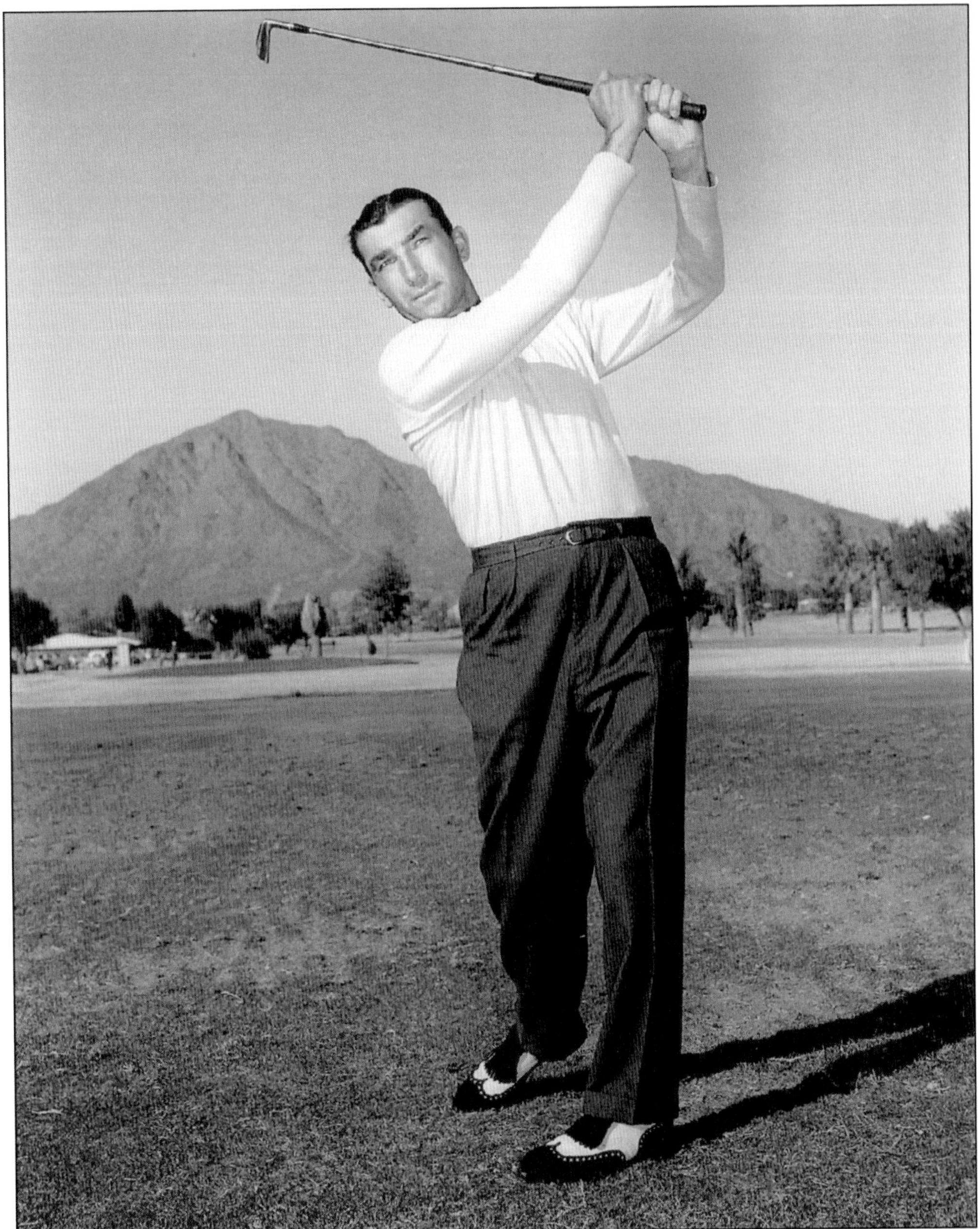

When he was growing up, Ed Furgol was in an accident that left him with a left arm several inches shorter than the right. This did not stop him from developing an unorthodox, but effective golf swing that resulted in his winning the US Open Championship in 1954 and being named the Professional Golfers Association Player of the Year. (NYMHS.)

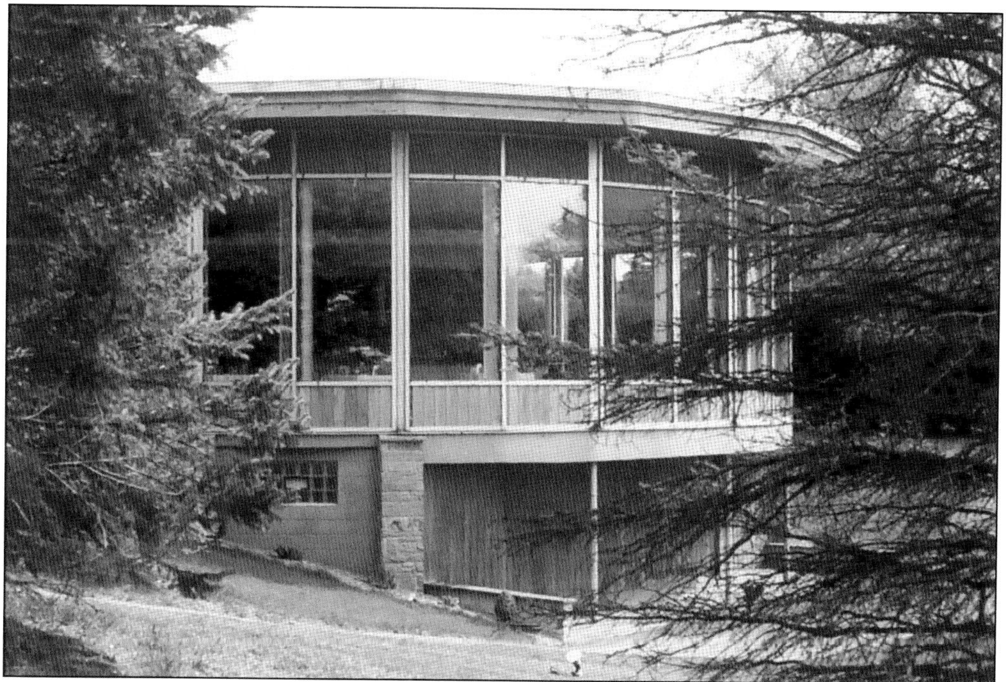

This glass-sided bar and restaurant overlooks a long fairway and green giving diners a perfect view of approach shots and green play. Its picturesque location makes Twin Ponds a favorite for corporate outings, parties, and individual relaxation. Several local leagues make the links their home course. (Authors' collection.)

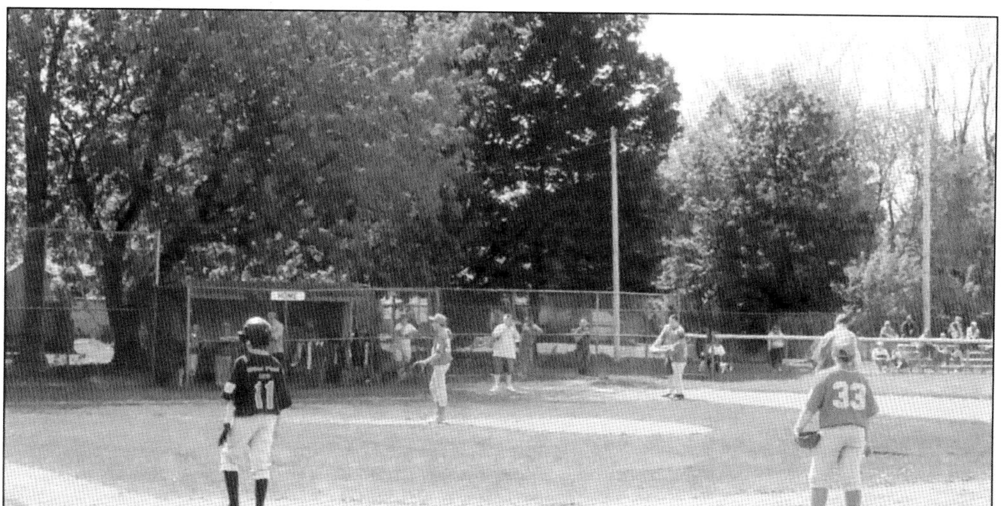

In 1953, Bator Post of the Polish Legion of American Veterans (PLAV) and the fire department cooperated in forming the New York Mills Little League, with each sponsoring one of the early teams along with Pezdeks-Quality Superette, the Marine Bar, and Consolidated Sheet Metal. The Little League field is located on land donated by the A.D. Juilliard Company between Clinton and Henderson Streets in the Upper Mills. (Authors' collection.)

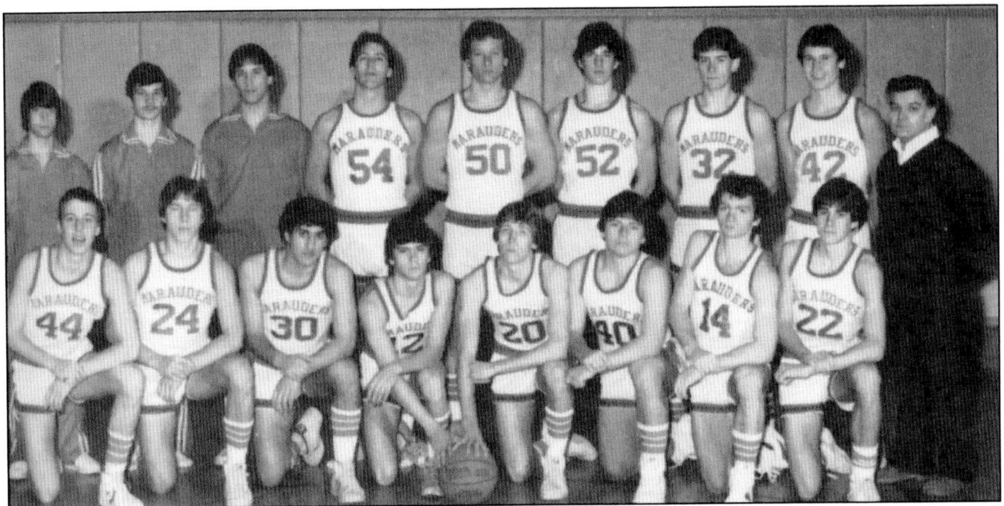

In 1982, the Marauders won their first New York State Class-D Championship. From left to right are (first row) David Comenale, Stan Rosinski, Jack DelGado, Anthony Abramowicz, Jim Humphrey, Trey Walewski, David Newman, and Christian Keator; (second row) manager Rich Bartoszek, manager John Marcinkiewicz, scorekeeper Mike Szczygiel, Chris Baron, Bob Kinville, Terry Inkawich, Frank Emmerich, Bill Herthum, and coach Mike Tomassetti. (NYMPS.)

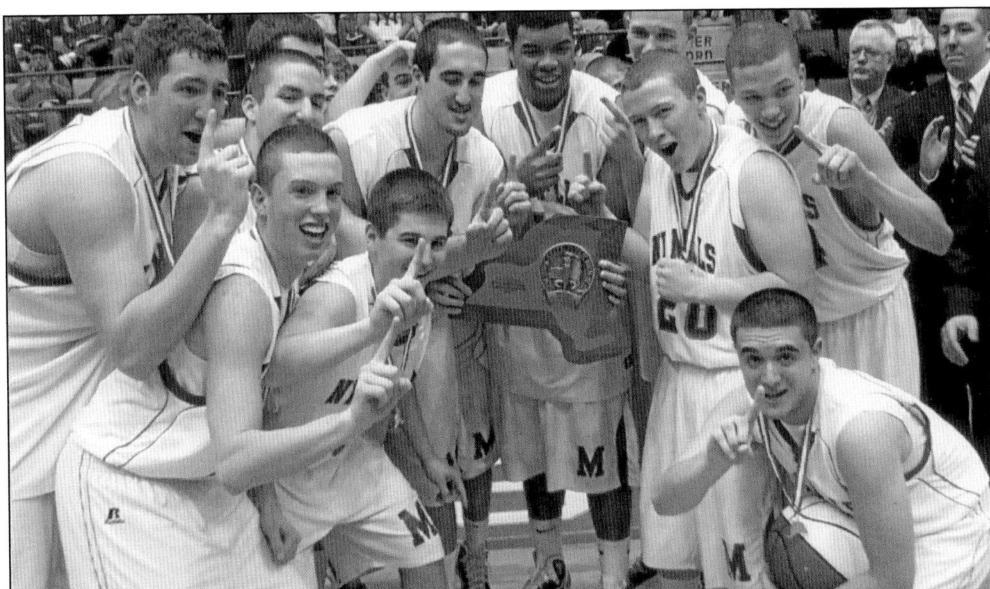

The 2011 New York State Class-D Championship team featured players Jeff Albright, Nick Eteinne, Jim Fellone, Shane Joswick, Jacob Kehrli, Rich Matrassi, Jon Ross, Fred Russ, Justin Stevener, Matt Welch, and Addison Wolanin; coaches Brian Adey, Mike Adey, Charlie Ferguson, Mike Keating, and John Stevener; managers George Albert, Nate Hartman, Cameron Jennings, Casey Rice, and Joe Topa. (NYMPS.)

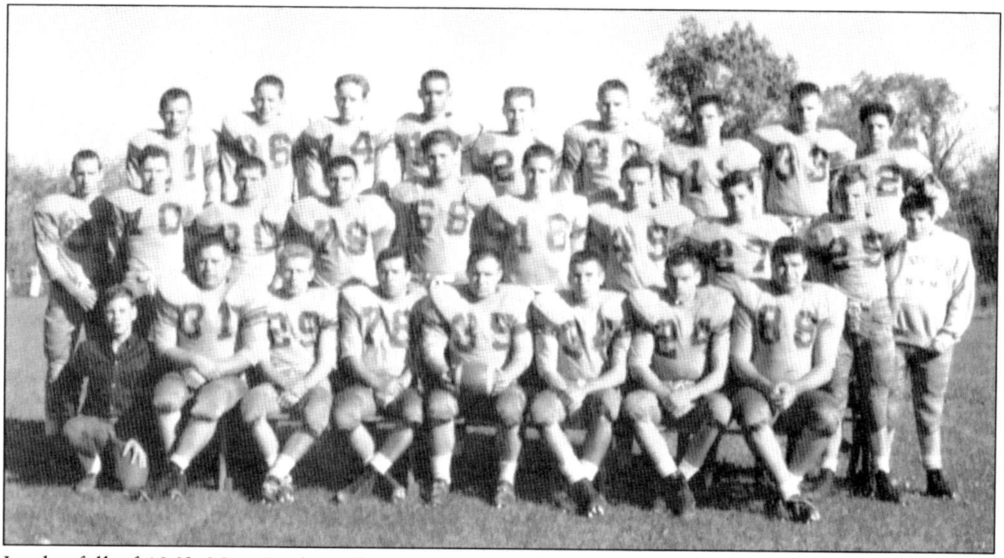

In the fall of 1963, New York Mills completed its first undefeated season in the tough Center State Conference, winning the league championship and gaining recognition throughout central New York. Its star running back, Terry Oakden, (No. 45, fourth from right in middle row), was later killed in Vietnam. (NYMPS.)

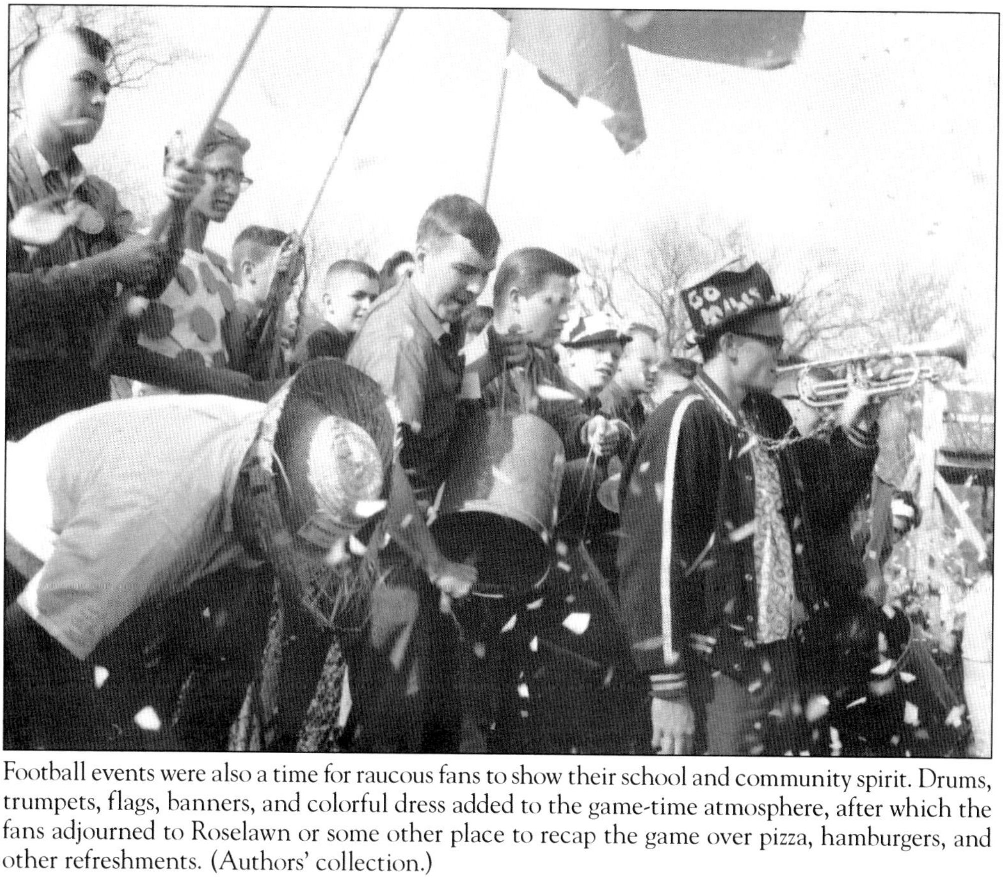

Football events were also a time for raucous fans to show their school and community spirit. Drums, trumpets, flags, banners, and colorful dress added to the game-time atmosphere, after which the fans adjourned to Roselawn or some other place to recap the game over pizza, hamburgers, and other refreshments. (Authors' collection.)

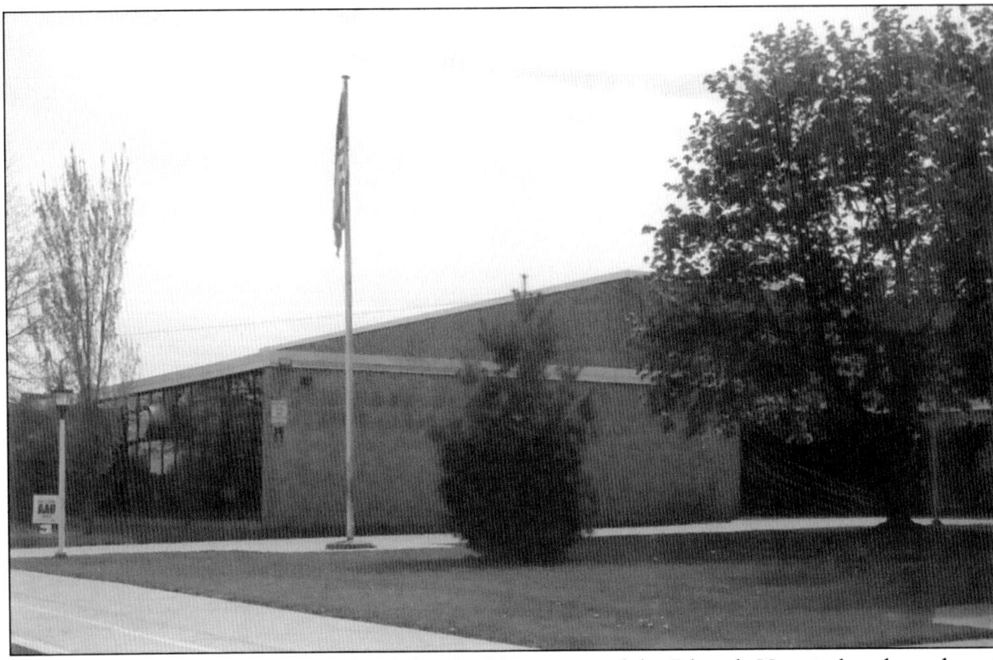

The new high school, located on land that had been part of the Rhoads Hospital and purchased from the government for $1, opened in 1958. Along with the new school, the district enjoyed the use of the Beekman Gymnasium, which was also ceded by the former hospital. Originally housing grades seven through 12, it was later expanded to include Kindergarten through 12th grade. (Authors' collection.)

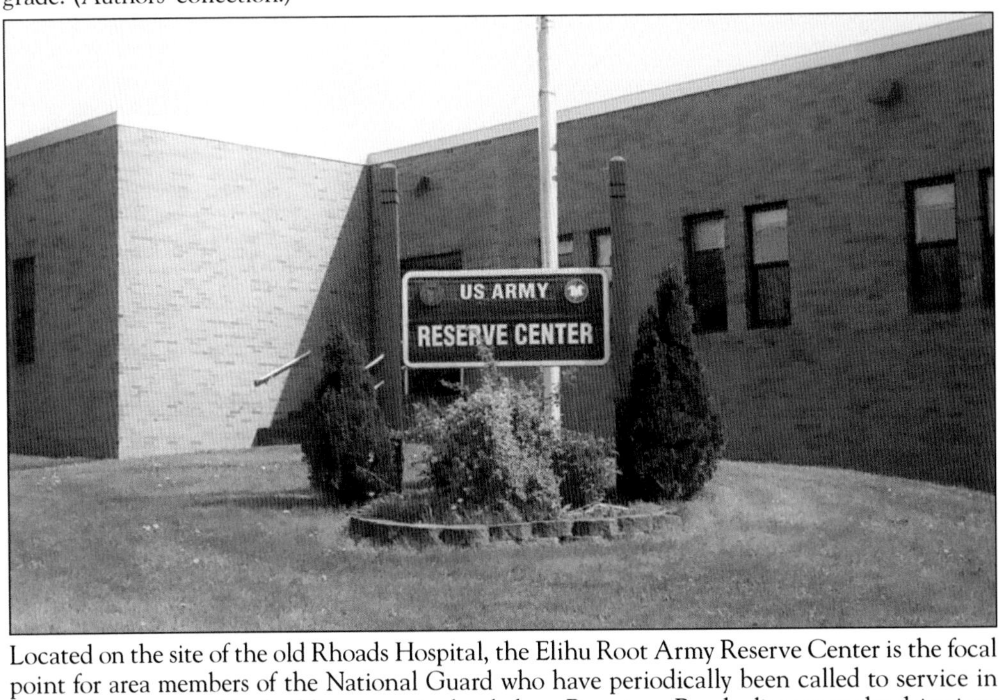

Located on the site of the old Rhoads Hospital, the Elihu Root Army Reserve Center is the focal point for area members of the National Guard who have periodically been called to service in times of emergency. The building occupies land along Burrstone Road adjacent to the drive into the high school (Authors' collection.)

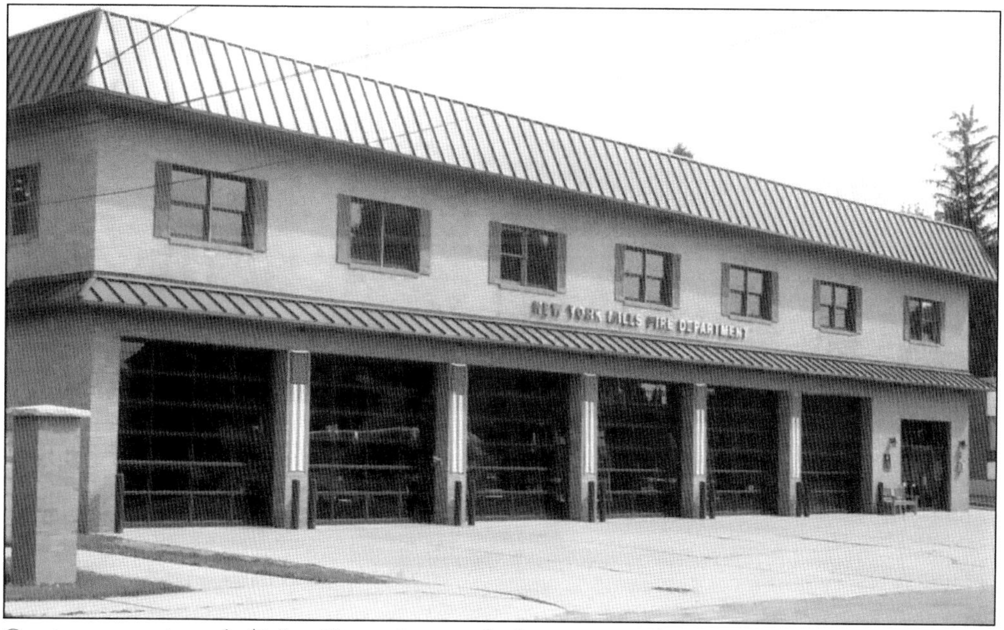

Costing approximately $1.5 million, the new firehouse opened in 1987 after demolition of the older structure that had served the community variously as a school, community building, fire house, and library for over a century. With six bays for apparatuses, it also included space for offices and meeting rooms on the second floor. (Authors' collection.)

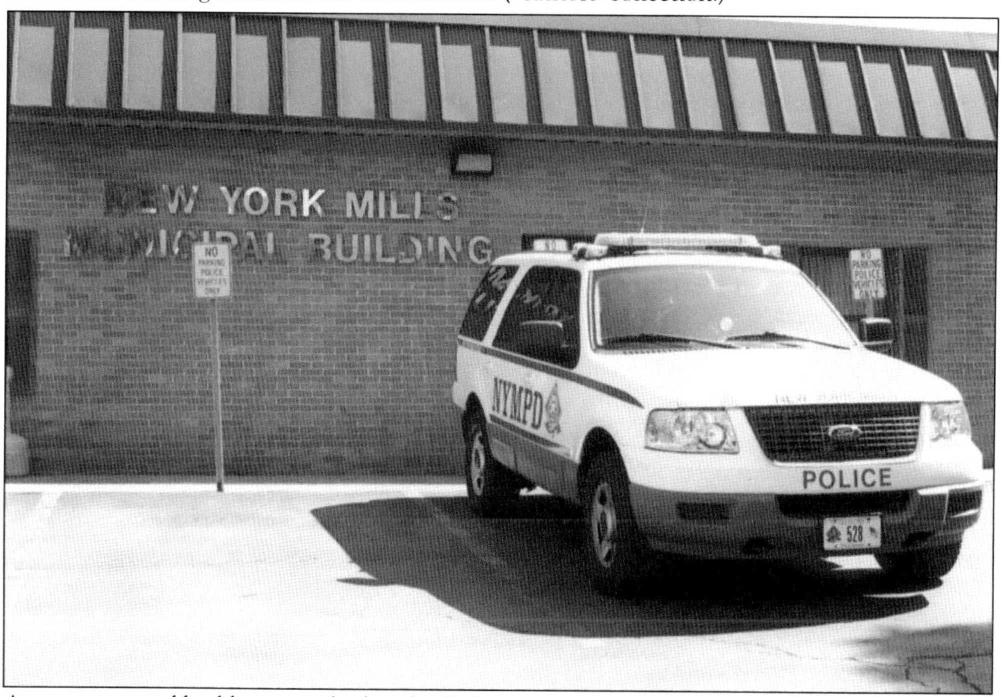

A new municipal building, attached to the firehouse, was dedicated in 1987. It contained a village boardroom, courtroom, police department offices and garage, village offices, and a general meeting room. Designed by Donald A. Engelbert, the facility contained 25,000 square feet at a cost of $1,042,280. (Authors' collection.)

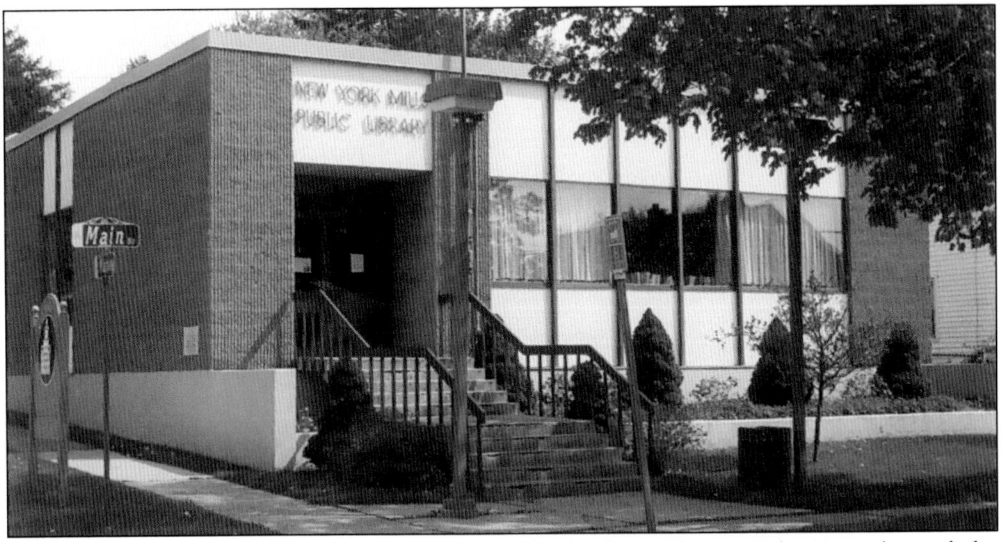

In 1973, a new library was dedicated under the leadership of Mayor John Pietryka and the president of the library board, Alice D. Finnegan. Village trustees were Charles Bosak, Walter Czajkowski, Joseph Hammond, and Thaddeus Kaczor. The architect was Jonza & Montary. The dedicatory address was given by Dr. James A. Smith of Syracuse University, a New York Mills native. (Authors' collection.)

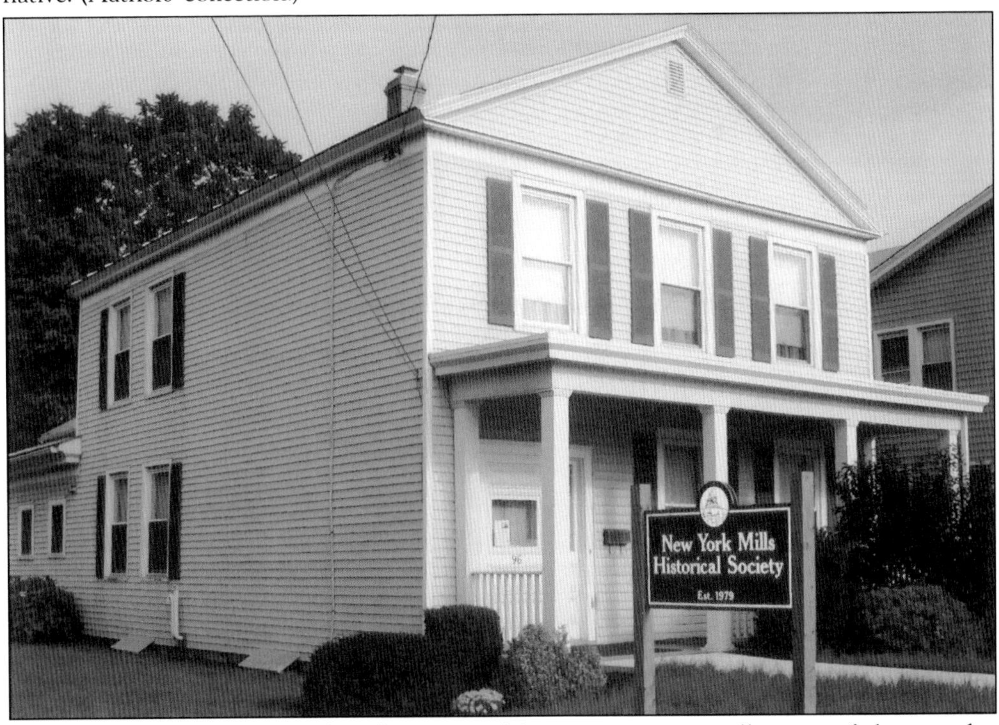

Founded in 1979, the New York Mills Historical Society maintains collections of photographs, artifacts, and memorabilia at 96 Main Street in the Lower Mills through the generosity of the Walter Karwowski family. Its first officers were Cheryl A. Pula, president; Eugene E. Dziedzic, vice president; Mildred Szarek, treasurer; Nancy Carey, secretary; and Mary Satterlee, public relations. (Authors' collection.)

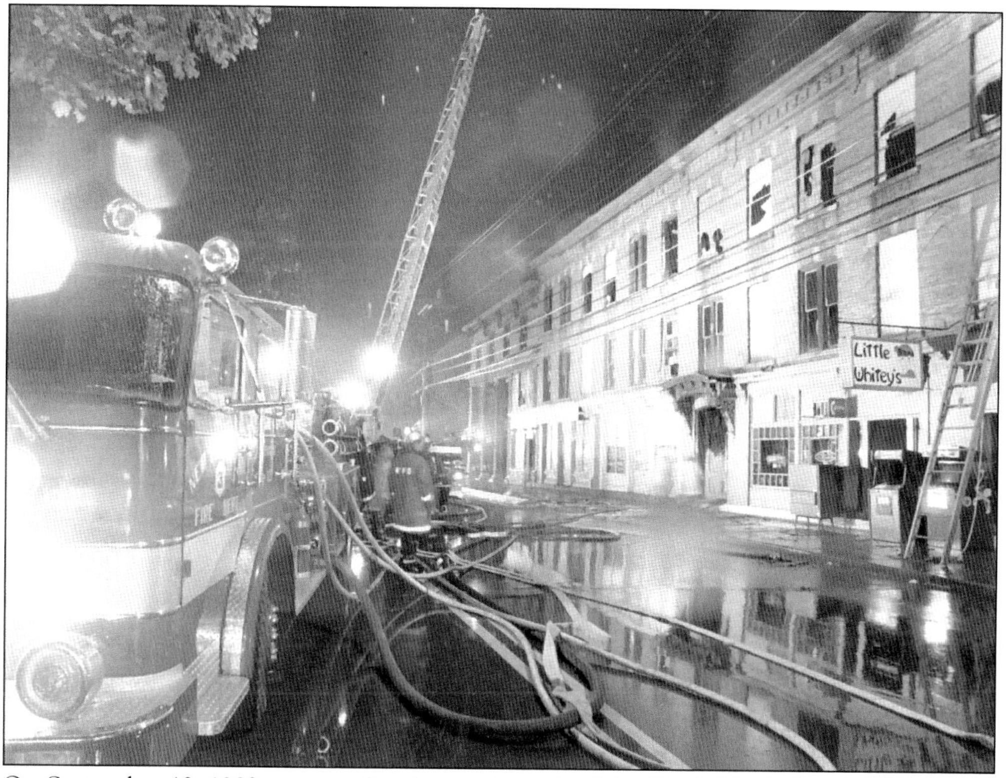

On September 10, 1983, a major fire damaged much of the Brick Block that housed several businesses. The extent of the fire made the building unsafe for further use. The potential cost of renovation resulted in it lying vacant for several years, a mere shell that reminded passersby of bygone years when it was a community hub. (NYMHS.)

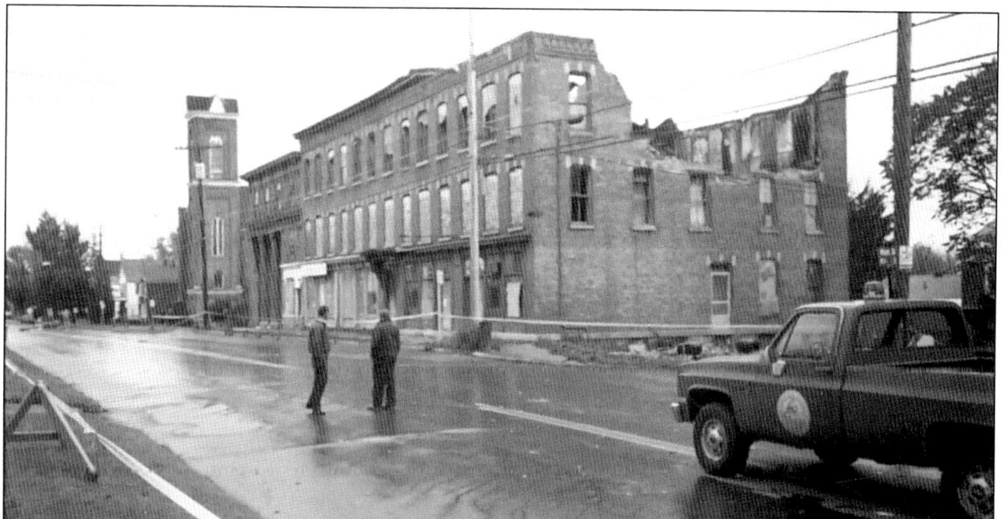

In 1990, a large portion of the fire-damaged Brick Block collapsed, including much of its roof, scattering bricks in all directions. Since the building was beyond repair—and a growing hazard—it was finally razed in October 1990. It was replaced with a large parking lot to serve the nearby library, village offices, and fire department. (Authors' collection.)

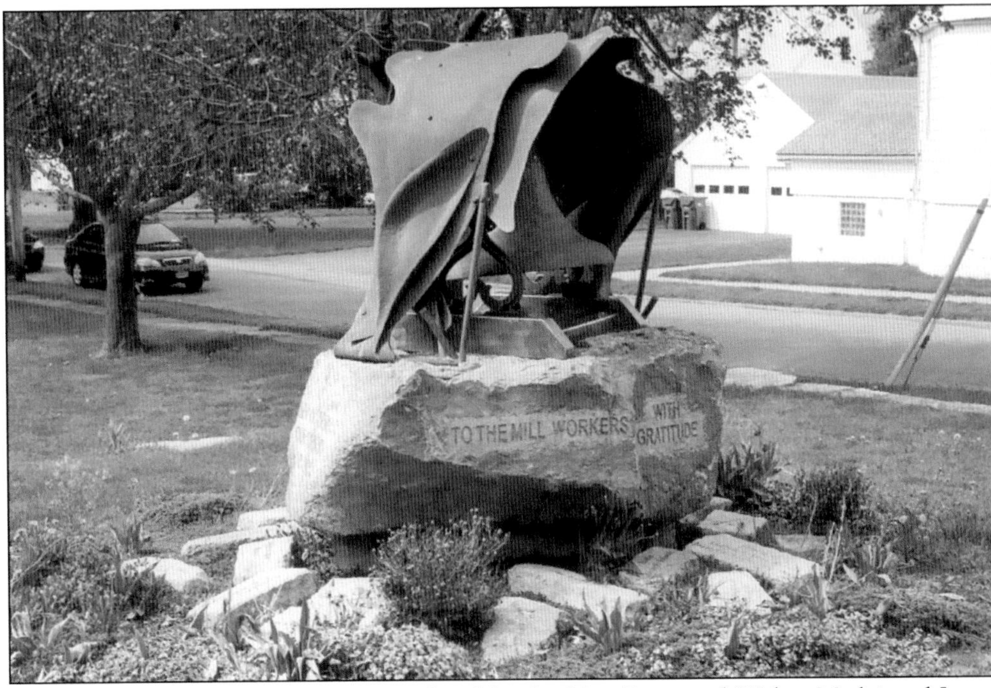

On July 8, 2000, a bell monument, sculpted by the Very Reverend Walter Madej and James McDermid, was unveiled in the village park. Designed to commemorate the contributions of the textile workers, it features a bell taken from the No. 2 mill that originally called the workers to the mills, announced the beginning and end of the lunch break, chimed at closing time, and proclaimed the evening curfew at 9:00 p.m. (Authors' collection.)

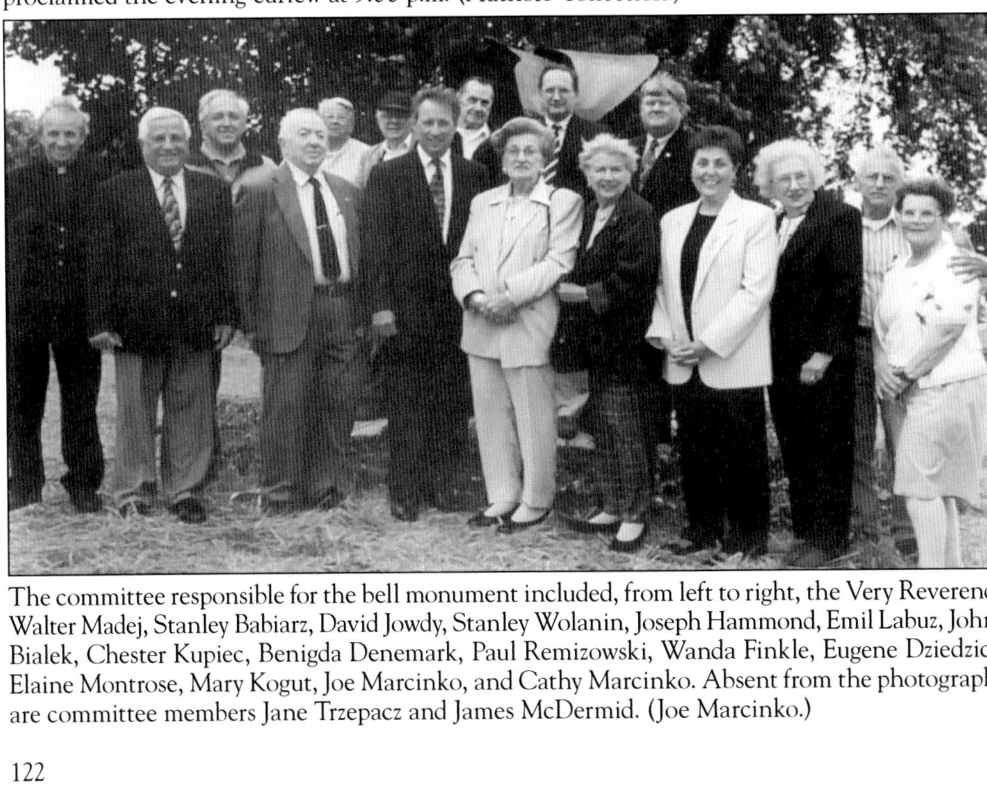

The committee responsible for the bell monument included, from left to right, the Very Reverend Walter Madej, Stanley Babiarz, David Jowdy, Stanley Wolanin, Joseph Hammond, Emil Labuz, John Bialek, Chester Kupiec, Benigda Denemark, Paul Remizowski, Wanda Finkle, Eugene Dziedzic, Elaine Montrose, Mary Kogut, Joe Marcinko, and Cathy Marcinko. Absent from the photograph are committee members Jane Trzepacz and James McDermid. (Joe Marcinko.)

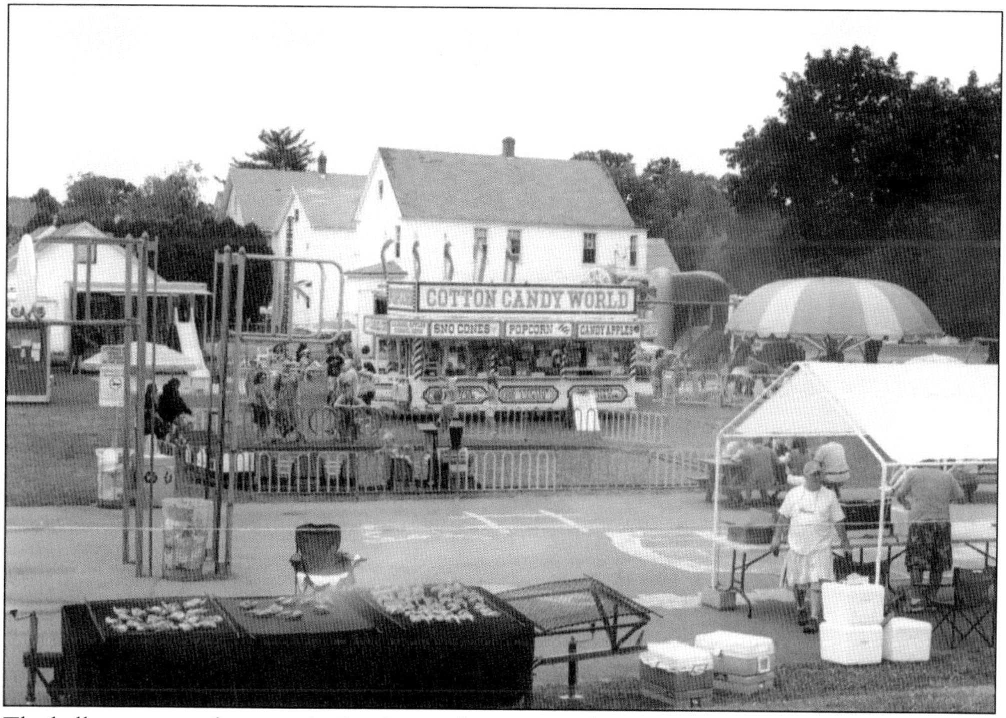

The bell monument became the focal point for an annual summer Bell Monument Festival, with games, carnival rides, and other entertainments, as well as a variety of cuisine. Former residents often returned to the village for a weekend of reliving old times, reconnecting with friends, and making new acquaintances. (C. Pula.)

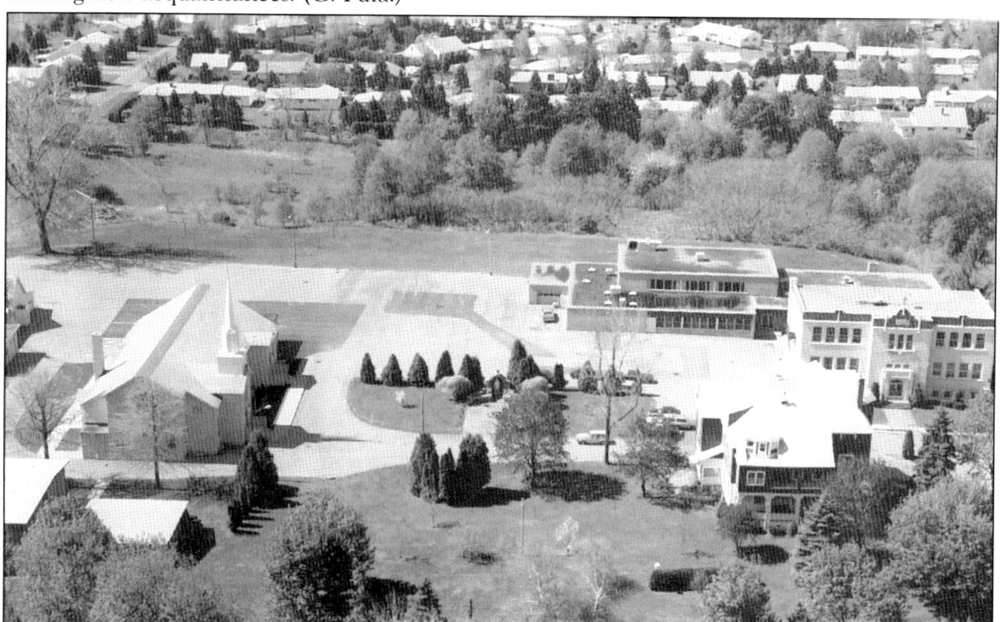

This aerial view of St. Mary's Parish in the 1960s shows the church to the left, the old school to the right, and the school addition, including a modern cafeteria immediately to the upper left of the old school. Below the old school is the rectory, the former Warner home. (Montrose.)

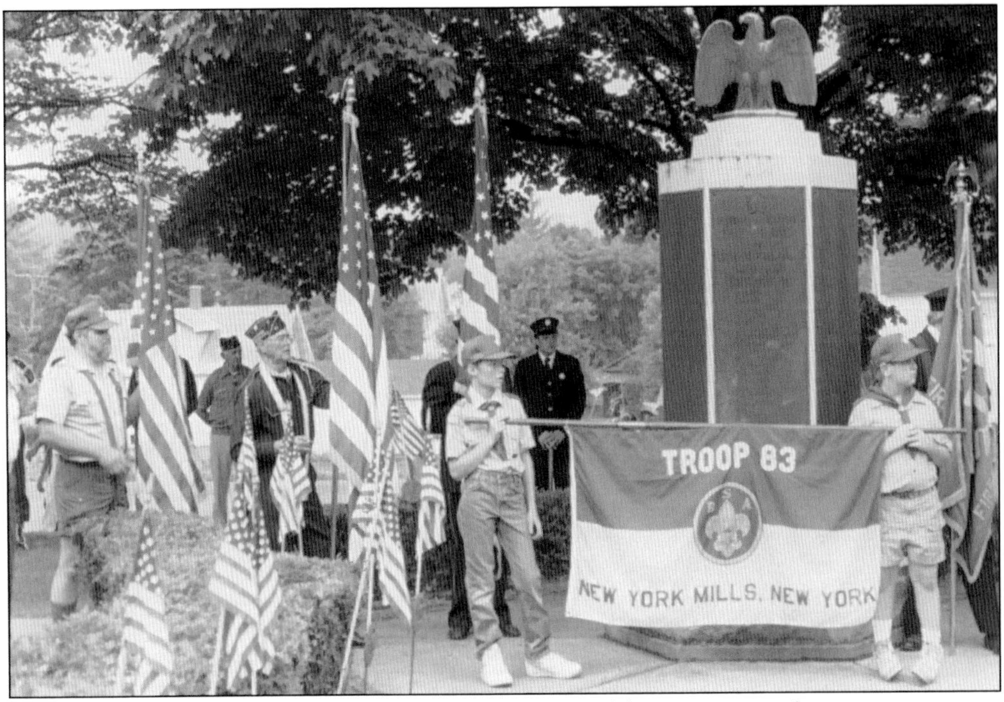

Boy Scouts, shown here participating in a Memorial Day celebration, is one of many organizations available to children in the village either through the school system, the local churches, or private groups. In this photograph, Scoutmaster Marty Pezdek appears at the far left of Troop 83. (Authors' collection.)

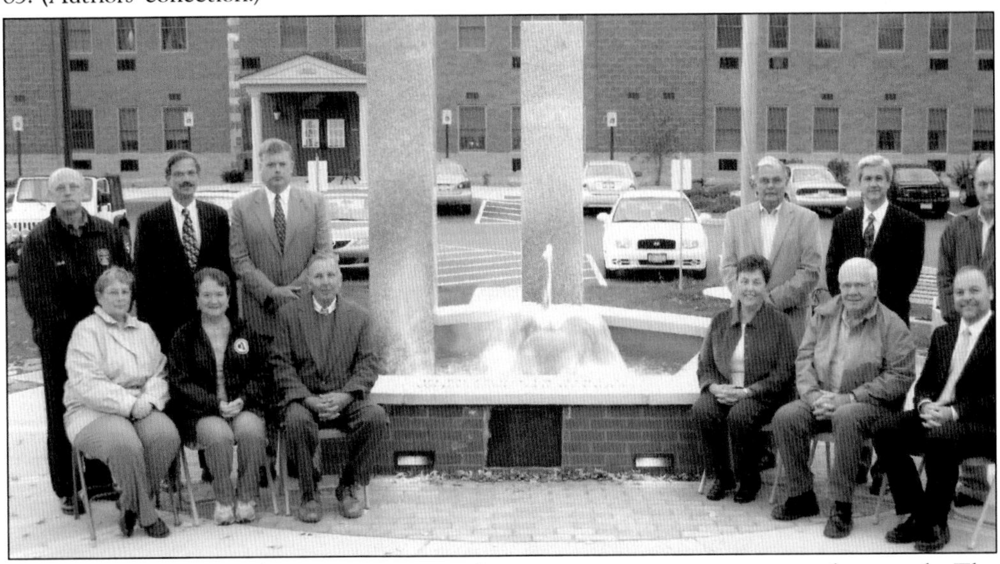

Following the 9/11 attack, a committee raised money to erect a monument to the tragedy. The 9/11 Memorial Committee included, from left to right, (first row, seated) Janet Griffiths, Nancy Czupryna, Stan Trzepacz, Elaine Montrose, Fred Griffiths, and William Borrill; (second row, standing) Rick Ulinski, Alan Swierczek, Robert Maciol, John Montrose, Mark Leogrande, and John Romano. Not present for the photograph was Swig Majka. The committee raised the entire amount necessary through private contributions. (Montrose.)

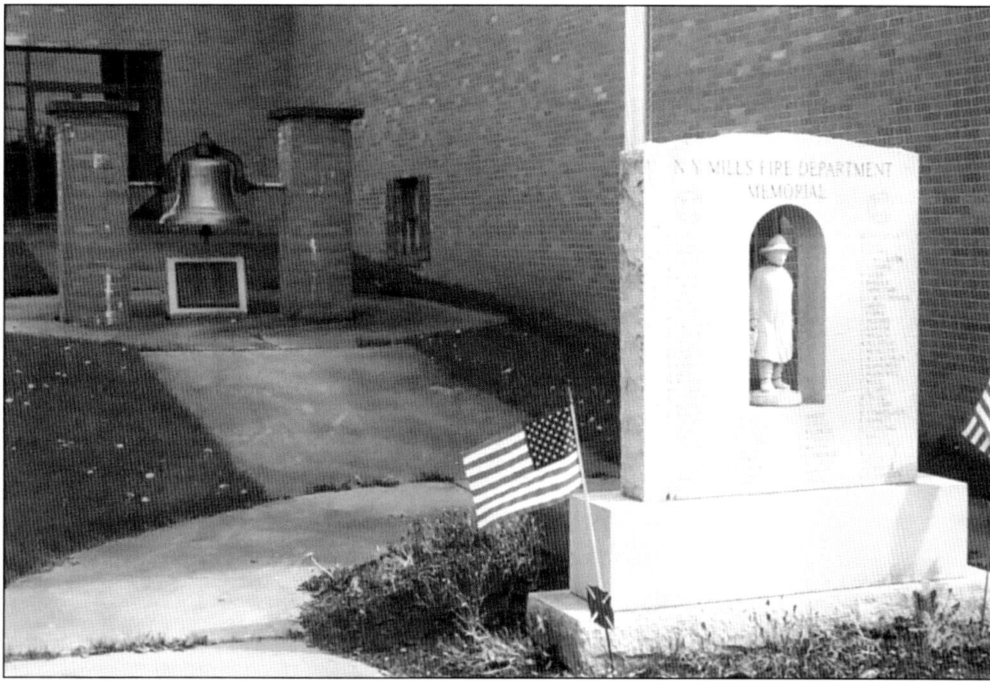

A memorial to the village's volunteer firemen was erected next to the firehouse including, in the background, a bell that once hung atop mill No. 3 where it rang the evening curfew until the factory burned in 1904. In the foreground is a monument that contains the names of deceased members of the New York Mills Fire Department. (Authors' collection.)

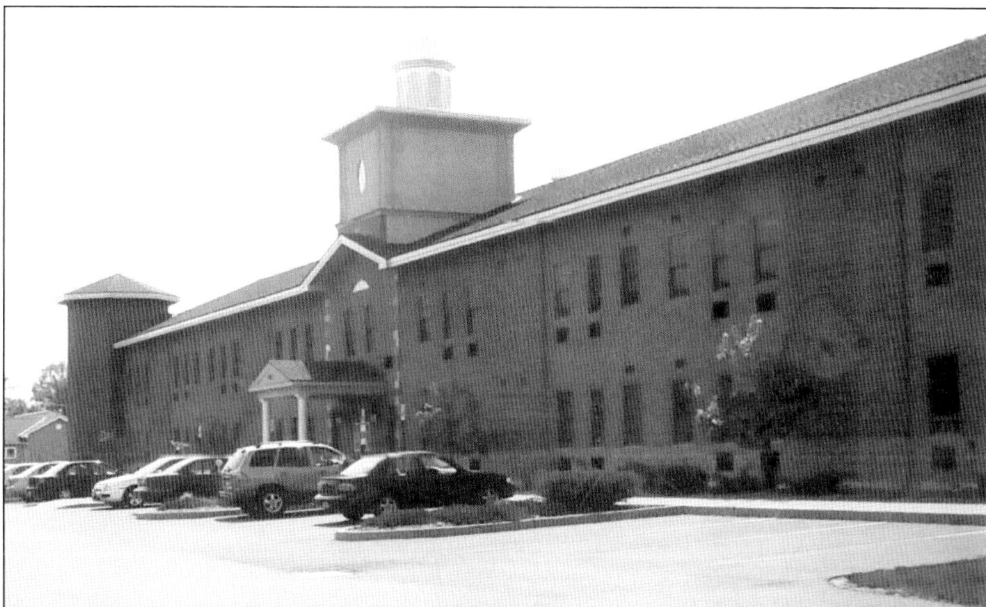

Opened in 2008 under Mayor Rob Maciol, the New York Mills Senior Center replaced the old, collapsed No. 2 mill. With 28 percent of the village population then over the age of 60 and another 11 percent aged 50 to 59, the facility was filled as soon as it opened. Its architecture is reminiscent of the textile era, and the red brick and contours evoke the missing mill. (Author's collection.)

BIBLIOGRAPHY

Davis, Susan G. "Women's Roles in a Company Town: New York Mills, 1900–1951." *New York Folklore*, Vol. 4, nos. 1–4 (1978): 21–34.

Hartman, Peter and Marc Tull. "Photographic Documentation of a Polish-American Community." *New York Folklore*, Vol. 4, nos. 1–4 (1978): 35–47.

Jareckie, Stephen B. "An Architectural Survey of New York Mills From 1808 to 1908." M.A. Thesis. Syracuse University, 1961.

Maziarz, Robert. "Polish Customs in New York Mills, N.Y." *New York Folklore*, Vol. 4, nos. 1–4 (1978): 302–307.

Pula, James S. and Eugene E. Dziedzic. *New York Mills: The Evolution of a Village*. Utica, New York: Ethnic Heritage Study Center, Utica College, 2004.

Pula, James S. and Eugene E. Dziedzic. *United We Stand: The Role of Polish Workers in the New York Mills Textile Strikes, 1912 and 1916*. New York: East European Monographs, Columbia University Press, 1990.

Savoie, Ernest J. "The New York Mills Company 1807–1914, A Study in Managerial Attitudes and Practices in Industrial Relations." M.A. Thesis. Cornell University, September 1955.

About the Organization

Located at 96 Main Street, the New York Mills Historical Society maintains exhibits, artifacts, photographs, archival materials, and other items related to the history, culture, and life of the village and its people (such as the label seen below). Materials, volunteers, and financial support are welcome. Regular meetings are held at 6:30 p.m. on the first Thursday of each month. For further information, contact the group by e-mail at nymhistoricalsociety@live.com, by telephone at 315-736-0489, or by mail at P.O. Box 151, New York Mills, New York, 13417.

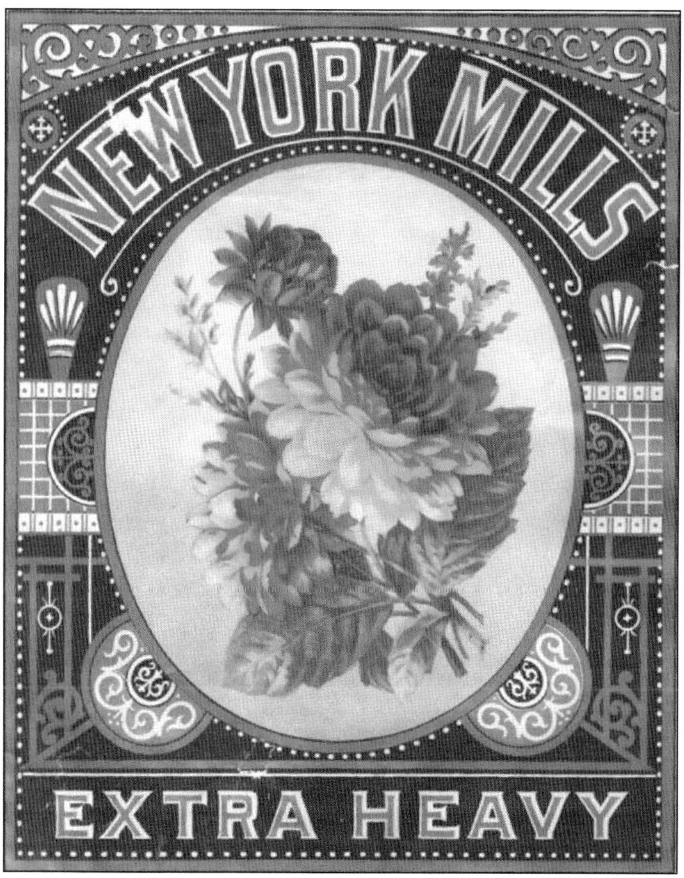

Consistent with our mission to preserve history on a local level, this book was printed in South Carolina on American-made paper and manufactured entirely in the United States. Products carrying the accredited Forest Stewardship Council (FSC) label are printed on 100 percent FSC-certified paper.